Lecture Notes in Mathematics

Edited by A. Dold and B. Eckmann

1131

Kondagunta Sundaresan
Srinivasa Swaminathan

Geometry and Nonlinear Analysis in Banach Spaces

Springer-Verlag
Berlin Heidelberg New York Tokyo

Authors

Kondagunta Sundaresan
Department of Mathematics, Cleveland State University
Cleveland, Ohio 44115, USA

Srinivasa Swaminathan
Department of Mathematics, Statistics and Computing Science,
Dalhousie University
Halifax, Nova Scotia B3H 4H8, Canada

Mathematics Subject Classification (1980): primary: 46B20, 58B10, 58C25
secondary: 41A65

ISBN 3-540-15237-7 Springer-Verlag Berlin Heidelberg New York Tokyo
ISBN 0-387-15237-7 Springer-Verlag New York Heidelberg Berlin Tokyo

Printed in Germany

Printing and binding: Beltz Offsetdruck, Hemsbach/Bergstr.
2146/3140-543210

CONTENTS

INTRODUCTION

This monograph is based in part on a series of lectures given by the first author at Cleveland State University and at Texas A & M University since 1980, and in part on the lectures given by the second author at Dalhousie University and the Australian National University since 1981. It concerns the development of certain topics in differential nonlinear analysis in infinite dimensional real Banach spaces. Our motivation derives from the rich and elegant theory of nonlinear analysis in finite dimensional setting and from the fact that the powerful theorems of Stone-Weierstrass, Whitney and Bernstein do not extend to infinite dimensional Banach spaces in general. There seems to be no comprehensive discussion available in the literature on the topics dealt with in these notes. During the last three decades many mathematicians have contributed to various problems on nonlinear analysis in Banach spaces which lie scattered in various journals. A substantial part of these contributions concern the unravelling of the geometric structure of infinite dimensional Banach spaces and the progress that has been made in applying these results to solve problems in nonlinear analysis on such spaces.

There are several excellent monographs on differential analysis in Banach spaces and the related theory of differentiable manifolds modelled on Banach spaces. In this connection we mention the books by Abraham and Robbin [1], Berger [5], Dieudonné [14], Lang [38] and Yamamuro [67] . While we refer to some of these contributions, we minimize overlap with the material in these works.

The nonlinear analysis in infinite dimensional Banach spaces dealt with in these notes is essentially concerned with functions of class C^k, i.e., k times continously Fréchet differentiable functions, for $k \geqslant 1$. In Chapter 1 we recall some basic definitions of smooth functions on open subsets of Banach spaces, some useful convexity properties of Banach spaces, the concept of finite representability, ultraproducts and a few inequalities concerning differentiable functions.

In Chapter 2 we provide a classification of Banach spaces based on the order of differentiability of the norm. We introduce C^k- , BF^k- and UF^k- smooth Banach spaces and discuss their interrelations. A differential characterization of Hilbert spaces modulo isomorphism and an isomorphic classification of superreflexive spaces are given.

In Chapter 3 we discuss various results of Bonic and Frampton [9], Torunczyk [62] and Wells [64] on smooth partitions of unity. We conclude the chapter with a nonlinear characterisation of superreflexive spaces and present a few applications of the characterization to differential analysis and approximation theory.

Chapter 4 is mainly concerned with the extensions of the well known theorems of Bernstein and Whitney and related theorems in the finite dimensional setting to infinite dimensional Banach spaces. The work of Aron and Prolla [4], Nachbin [44], Kurzweil [37] and Restrapo [53] on approximation by differentiable and analytic functions, is discussed. Some recent results of Moulis [43] and Heble [25] on simultaneous approximations by differentiable functions on certain smooth Banach spaces are stated.

The volume concludes with an appendix on differentiable manifolds modelled on Banach spaces, dealing with the diffeomorphism and embedding theorems of Bessaga [6] and Eells-Elworthy [21], and the generalisation of Palais of Morse's theorem on the behaviour of a smooth function in the neighborhood of a nondegenerate critical point to infinite dimensional Banach spaces.

The first author acknowledges his gratitude to Victor Klee for suggesting certain problems concerning C^k-norms on Banach spaces and to him and R.R.Phelps for valuable discussions. The authors express their thanks to Richard Aron for carefully reading a first version of the chapters and making valuable suggestions. They are grateful to the referee for comments and suggestions for improvement. Further they acknowledge their gratitude for the facilities provided by Cleveland State University and Dalhousie University in carrying out this project. The first author wishes to thank Elton Lacey for providing an opportunity to deliver some lectures on the topics dealt with in these notes in the Banach Space Seminars at Texas A & M University during the Spring semester of 1981. The second author acknowledges support of a grant (A 5615) from NSERC(Canada) and also takes this opportunity to thank the Research School of the Australian National University, Canberra, for facilities afforded in 1984.

Finally the authors express their appreciation and thanks to the Editors of the LECTURE NOTES Series for their advice and encouragement.

Chapter 1

Basic Definitions and Geometric Properties

In this introductory chapter we introduce the basic notations and definitions, and recall certain geometric properties essential to our discussion.

All Banach spaces considered here are over the real field R. If $(E, \| \ \|)$ is a Banach space then the conjugate of E is denoted by E^*, and $\| \ \|^*$ is the norm conjugate of the norm $\| \ \|$. If E, F are two Banach spaces then L(E,F) is the Banach space of continuous linear operators on E into F with the supremum norm. The Banach space of continuous k-linear operators T^k on E into F is denoted by $B^k(E,F)$ with the norm defined by

$$\|T^k\| = \sup\{\|T^k(x_1,x_2,\ldots,x_k)\| \mid x_i \in E, \quad \|x_i\| \le 1\}$$

where $T^k \in B^k(E,F)$. When F is the one dimensional space R the space $B^k(E,F)$ is simply denoted by $B^k(E)$.

1.1 Some Geometric Properties of Banach Spaces

1.1.1 Convexity and smoothness properties

A Banach space $(E, \| \ \|)$ is said to be strictly convex if $x,y \in E$, $x \neq y$, and $\|x\| = \|y\| = 1$ imply $\|\frac{x+y}{2}\| < 1$. E is uniformly convex, if given any $\epsilon > 0$ there is a $\delta(\epsilon) > 0$, such that $\|x\| = \|y\| = 1$, $\|x-y\| \ge \epsilon$ imply $\|\frac{x+y}{2}\| \le 1 - \delta(\epsilon)$. If $x \in E$, $\|x\| = 1$, E is said to be smooth at x, if there is a unique $\ell_x \in E^*$ with $\|\ell_x\| = 1$, such that $\ell_x(x) = 1$. $\ell_x \in E^*$ is called the support functional for the unit ball U of E at x and $\ell_x^{-1}(1)$ is the hyperplane of support for U at x. It is well known that U is smooth at x iff

(a) $$\lim_{t \to 0} \frac{||x+ty|| - ||x||}{t} = g_x(y) \text{ exists for all } y \in E.$$

Further the limit in (a) exists iff the limit with x replaced by λx, $\lambda \neq 0$ exists. Further $g_{\lambda x} = (\text{sign } \lambda) g_x$. The norm functional on E is said to be differentiable (Fréchet) if for each $x \neq 0$ there exists a linear functional $\ell_x \in E^*$ such that

(b) $$\lim_{||h|| \to 0} \frac{||x+h|| - ||x|| - \ell_x(h)}{||h||} = 0.$$

E is said to be uniformly smooth if the limit in (a) exists uniformly for all (x,y), $||x|| = 1 = ||y||$. It is known that E is uniformly smooth if the limit in (b) is uniformly attained for all x, $||x|| = 1$.

In the following remark we summarize some well known results concerning the preceeding properties.

1.1.2 Remark

It is well known that if the limit $\lim_{t \to 0} \frac{||x+ty|| - ||x||}{t} = g_x(y)$ exists at some $x \neq 0$, for all $y \in E$, then $\ell_x \in E^*$, and $||\ell_x|| = 1$. Further if the dual of a Banach space E is smooth (strictly convex) iff E is strictly convex (smooth). A Banach space $(E, || \ ||)$ is uniformly convex (uniformly smooth) iff E^* is uniformly smooth (uniformly convex). For these results see, Day [11].

1.2 Finite representation of a Banach space

A normed linear space E is said to be finitely represented in another normed linear space F in symbols $E \ll F$, if for each $\epsilon > 0$, and each finite dimensional subspace E_0 of E there exists a subspace F_0 of F, depending on E_0 and ϵ, such that there is an isomorphism T on E_0 onto F_0 satisfying $\|T\| \; \|T^{-1}\| \leq 1 + \epsilon$.

A useful tool in the theory of finite representation is the concept of ultraproduct of a normed linear space. Let S be an infinite set, and $\mathcal{U}$ be a nontrivial (free) ultrafilter on S. If f is a bounded real valued function on S, then $\lim_{\mathcal{U}} f$ is by definition the number, $\sup[\lambda \mid \{t \in x, f(t) > \lambda\} \in \mathcal{U}]$. If $(E, \| \; \|)$ is a normed linear space, and f is a bounded E-valued function on S, let $|f| = \lim_{\mathcal{U}} \|f(t)\|$. It is verified that $| \; |$ is a seminorm on the vector space V of bounded E-valued functions on S. The quotient space of E modulo the kernel of the seminorm $| \; |$, equipped with the quotient norm is called the ultrapower of E with respect to the pair $(S, \mathcal{U})$ and we denote this here by $E(S, \mathcal{U})$.

In the discussion to follow the equivalence class determined by a bounded function $f:S \to E$ is simply denoted by $\tilde{f}$. Further for clarity, we write sometimes $\{f(s)\} \in \tilde{f}$.

We summarize here several useful facts concerning ultrapowers. For a detailed account concerning ultrapowers and their ramifications in the structure theory of Banach spaces, we refer to Stern [56], and Krivine [34].

1.2.1 <u>Proposition</u>

The ultrapower $E(S, \mathcal{U})$ of a Banach space E is a Banach space.

For subsequent use in our discussions we recall some basic concepts from [34].

If E is a Banach space, and $x_1, \ldots, x_n \in E$, let $\Phi: R^n \to R$ be defined by $\Phi(\lambda_1, \lambda_2, \ldots, \lambda_n) = \|\sum_{i=1}^{n} \lambda_i x_i\|$. Φ is called a n-type associated with $x_1, x_2, \ldots, x_n$. Since Φ is absolutely homeogeneous, Φ is determined by its values on the set $S_\infty^n = \{(\lambda_1, \lambda_2, \ldots, \lambda_n) \in R^n \mid \sup_{1 \leq i \leq n} |\lambda_i| = 1\}$. The set of all functions Φ associated with n-tuples of points $\{x_i\}_{i=1}^n \subset E$ could be equipped with the topology τ of uniform convergence on S_∞^n.

1.2.2 Lemma

Let E be a Banach space, and $E(S,\mathcal{U})$ be an ultrapower of E. Let $\{\tilde{f}_i\}_{i=1}^{n} \subset E(S,\mathcal{U})$ and $\{f_i(s)\}_{s\in S} \in f_i$, $1 \le i \le n$. Let Φ_s, Φ be the n-types of $\{f_i(s)\}_1^n$, and $\{f_i\}_1^n$ respectively. Then $\Phi_s \to \Phi$ following the ultrafilter $\mathcal{U}$ in the topology τ of n-types.

Proof

Since the ranges of f_i are bounded subsets of E, there is a number $M > 0$ such that

$$\|f_i(s)\| \le M, \quad |||f_i||| \le M, \quad 1 \le i \le n, \quad s \in S.$$

From the definition of ultrapowers it follows that

(1) $\lim \Phi_s(\lambda_1,\lambda_2,\ldots,\lambda_n) = \Phi(\lambda_1,\lambda_2,\ldots,\lambda_n)$ for $(\lambda_1,\lambda_2,\ldots,\lambda_n) \in R^n$.

Further

$$|\Phi_S(\lambda_1,\lambda_2,\ldots,\lambda_n) - \Phi_S(\lambda_1^1,\lambda_2^1,\ldots,\lambda_n^1)|$$
$$\le \|\Sigma(\lambda_i - \lambda_i^1) f_i(s)\|$$
$$\le k \sum_{i=1}^{n} |\lambda_i - \lambda_i^1|.$$

Thus $\{\Phi_s\}_{s\in S}$ is an equicontinuous family of functions on the compact set S_∞^n. Thus by Arzela's theorem it follows that $\lim_{\mathcal{U}} \Phi_s(\lambda_1,\lambda_2,\ldots,\lambda_n) = \Psi(\lambda_1,\ldots,\lambda_n)$ for some continuous function Ψ on $S_\infty^n \to R$, uniformly over S_∞^n. From (1) $\Psi = \Phi$. Thus $\lim \Phi_s = \Phi$ uniformly over S_∞^n.

The main theorem which establishes the importance of the concept of ultrapowers of Banach spaces in the theory of finite representation is the following. Since the proof illustrates a technique we provide the details here.

1.2.3 Theorem

Let E, F be two Banach spaces. Then $E \ll F$ iff E is isometric with a subspace of an ultrapower of F.

Proof

Let $E \ll F$. Let $\mathcal{F}$ be the set of all finite dimensional subspaces of E. Consider the free ultrafilter $\mathcal{U}_1$ generated by the tails $\{L \in \mathcal{F} \mid L \supset L_0\}$ where L_0 is an arbitrary subspace of E in $\mathcal{F}$. Let I be the set of positive integers, and $\mathcal{U}_2$ be a free ultrafilter of I. Let $\mathcal{U}$ be the ultrafilter on $S = \mathcal{F} \times I$ generated by the product $\mathcal{U}_1 \times \mathcal{U}_2$. Let $T:E \to F(S,\mathcal{U})$ be defined by setting $T_x = T_L^n(x)$ if $x \in L$, and $Tx = 0$ otherwise, where T_L^n is an isomorphism on $L \to F$ such that $\|T_L^n\| \cdot \|(T_L^n)^{-1}\| \leq 1 + \frac{1}{n}$. Since $\|Tx\| = \lim_{(L,n)\in\mathcal{U}} \|T_L^n(x)\|$, it follows that $\|Tx\| = \|x\|$. We complete the proof by showing that the ultrapower $F(S, \mathcal{U}) << F$. Let G be a finite dimensional subspace of $F(S, \mathcal{U})$, and $\{\tilde{f}_i\}_{i=1}^n$ be a basis for G. Let $\{f_i(s)\}_{s\in S} \in \tilde{f}_i$, $1 \leq i \leq n$. Since $|||\Sigma\lambda_i\tilde{f}_i|||$, $\sup_{1\leq i\leq n} |\lambda_i|$, are norms on R^n, it follows that there exists a constant k such that

$$k\ |||\Sigma\lambda_i\tilde{f}_i||| \geq \sup_{1\leq i\leq n} |\lambda_i|, \text{ for all points } (\lambda_1,\lambda_2,\ldots,\lambda_n) \in R^n.$$

Now if $\epsilon > 0$ from the preceeding lemma it follows that there is a set $Q \subset S$, $Q \in \mathcal{U}$ such that $\left|\ |||\Sigma\lambda_i f_i||| - \|\Sigma\lambda_i f_i(s)\|\ \right| \leq \epsilon$ for all $s \in Q$, and for all $(\lambda_1,\lambda_2,\ldots,\lambda_n) \in R^n$. Let $T:G \to F$ be defined by $T(\Sigma\lambda_i\tilde{f}_i) = \Sigma\lambda_i f_i(s_0)$ where s_0 is a fixed point in Q. If $T(G) = M$, it follows that T is an isomorphism on G onto M, and $\|T\| \cdot \|T^{-1}\| \leq (1 + k\epsilon)/(1 - k\epsilon)$.

1.2.4 <u>Definition</u>

A Banach space E is superreflexive if every Banach space F finitely represented in E is reflexive, James [28]. For an account of superreflexive spaces see VanDulst [63] and Diestel [12].

1.2.5 <u>Theorem</u> [James and Enflo]

The following statements concerning a Banach space E are equivalent:

(a) E is superreflexive.

(b) E is isomorphic to a uniformly smooth (uniformly convex) Banach space F.

(c) E is isomorphic to a Banach space F which is uniformly

smooth and uniformly convex.

For a proof of the theorem we refer to [63].

1.3 Multilinear forms and differential concepts in Banach spaces

1.3.1 Remarks on multilinear forms

If T^n is a continuous n-linear form on a Banach space E into R, we already noted in 1.1 that the norm of T^n is given by

$$\sup_{\substack{\|x_i\|=1 \\ 1\le i\le n}} |T^n(x_1,x_2,\ldots,x_n)|.$$

If T^n is also symmetric then by the polarization identity, Alexiewicz and Orlicz [2], it follows that

$$T^n(x_1,x_2,\ldots,x_n) = \frac{1}{n!} \sum_{\epsilon_1,\epsilon_2,\ldots,\epsilon_n=0}^{1} (-1)^{n-\sum_{i=1}^{n}\epsilon_i}\, T^n\Big(\sum_{i=1}^{n}\epsilon_i x_i\Big)^{(n)},$$

where $T^n(h)^{(n)} = T^n(h,h,\ldots,h)$. Hence it follows that

$$\|T^n\| \le \frac{2^n n^n}{n!} \sup_{\|x\|=1} |T^n(x,x,\ldots,x)| \le \frac{2^n n^2}{n!} \|T^n\| \quad \ldots \quad \text{(A)}$$

The definition of a polynomial in a real variable may be extended to Banach spaces. If T^n is a continuous symmetric n-linear form on a Banach space E, then $T^n(x,x,\ldots,x)$, usually denoted by $T^n(x^{(n)})$, is called a homogeneous polynomial of degree n. A polynomial of degree n on a Banach space E is a function $P(x) = \sum_{i=0}^{n} T^i(x^{(i)})$, where $T^i(x^{(i)})$ is a homogeneous polynomial of degree i, $1 \le i \le n$.

1.3.2 Definition

If U is an open subset of a Banach space E and f is a function on U into a Banach space F then f is said to be differentiable (Fréchet) at $x \in U$ if there is a continuous linear operator T_x on $E \to F$ such that

$$\lim_{\|h\|\to 0} \frac{\|f(x+h)-f(x)-T_x(h)\|}{\|h\|} = 0.$$

T_x is called the derivative of f at x. If f is differentiable at all points $x \in U$, we say that f is differentiable in U, and the map $x \to T_x$ on U into L(E,F) is called the differential of f, and is denoted by Df. If Df is a continuous map then f is called a C^1-function. f is said to be k-times differentiable in U, if $D^{k-1}f$ exists and the map $D^{k-1}f: U \to B^{k-1}(E,F)$ is once differentiable, and in this case the kth differential $D^k f$ is a function on U into $B^k(E,F)$, the Banach space of continuous k-linear operators on E into F. If $D^k f$ is continuous we say f is of class C^k. Further we note that $D^k f(x)$ is a symmetric k-linear operator on E into F. The usual form of Taylor's theorem in the finite dimensional calculus extends to infinite dimensions equally well. Thus if f is a C^k map on U into F, and $x \in U$, then if $[x, x+h] \subset U$,

$$\text{(A)}\quad f(x+h) = f(x) + T_x^1(h) + T_x^2(h^{(2)}) + \ldots + T_x^k(h^{(k)}) + \theta_x(h)$$

where (1) T_x^i are continuous symmetric i-linear maps on E into F, (2) the maps $x \to T_x^i$ are continuous on U into $B^i(E,F)$, and

$$\text{(3)}\quad \lim_{||h|| \to 0} \frac{||\theta_x(h)||}{||h||^k} = 0.$$

In (A) above, $T_x^i(h^{(i)}) = \frac{1}{i!} f^i(x) \cdot h^{(i)}$, $1 \le i \le k$ where $f^i(x)$ is the ith derivative at x. Further sometimes it is useful to use the following form (Lagrange formula) of (A).

$$f(x+h) = f(x) + \sum_{i=1}^{k-1} \frac{1}{i!} f^i(x)(h^{(i)}) + \left(\int_0^1 \frac{(1-s)^{k-1}}{(k-1)!} f^k(x+sh)\,ds\right)h^{(k)}.$$

Further we have the following converse of Taylor's theorem. If $f: U \to F$ is a continuous mapping such that for each $x \in U$,

$$f(x+h) = f(x) + \sum_{i=1}^{k} T_x^i(h^{(i)}) + \theta_x(h)$$

where the maps $x \to T_x^i$ are continuous on U into the space $B^i(E,F)$, with T_x^i symmetric for all $x \in U$, $1 \le i \le k$, and if $\frac{||\theta_x(h)||}{||h||^k} \to 0$ as $||h|| \to 0$, then f is a C^k-function on $U \to F$. See [45].

In differential analysis in infinite dimensional Banach spaces it is extremely useful to introduce weaker as well as stronger differentiability of a function than the customary C^k-differentiability as already done here above. We introduce few such concepts here, as they are useful in the discussion to follow.

1.3.3 Definition

Let U be an open subset of a Banach space E and f be a continuous mapping on U into a Banach space F. If $x \in U$, then f is said to be k-times directionally differentiable at x if there are continuous symmetric i-linear transformations T_x^i, $1 \le i \le k$ on E into F such that for each $h \in E$

$$(1) \quad f(x+th) = f(x) + \sum_{i=1}^{k} t^i T_x^i(h^{(i)}) + \theta_x(th) \quad \text{where} \quad \frac{\|\theta_x(th)\|}{|t|^k} \to 0 \text{ as } t \to 0.$$

It is verified that the i-linear transformations $\{T_x^i\}$, $1 \le i \le k$, are uniquely determined. f is said to be k-times Fréchet differentiable (F-differentiable) if the limit in (1) is uniform over the unit sphere, $S_E = \{h \mid h \in E, \|h\| = 1\}$ of E i.e. given $\epsilon > 0$ there is a $\delta > 0$ independent of $h \in S_E$ such that $\|\theta_x(th)\| \le \epsilon |t|^k$, if $|t| < \delta$. f is said to be k-times directionally (k-times Fréchet) differentiable over the set U if f is k-times directionally (k-times Fréchet) differentiable over U. f is said to be uniformly k-times Fréchet differentiable over a set $P \subset U$ if for each $x \in P$

$$(2) \qquad f(x+h) = f(x) + \sum_{i=1}^{k} T_x^i(h^{(i)}) + \theta_x(h) \quad \text{where} \quad \{T_x^i\},\ 1 \le i \le k$$

are as described above, and $\lim_{\|h\| \to 0} \frac{\|\theta_x(h)\|}{\|h\|^k} = 0$ uniformly for $x \in P$. If f is k-times Fréchet differentiable at all $x \in P$, then f is said to be boundedly k-times Fréchet differentiable over P if $\sup_{x \in P} \|T_x^k\| < \infty$.

The following proposition follows from the definitions, and Taylor's theorem and its converse.

1.3.4 Proposition

If U is an open subset of a Banach space E, and f is k-times

Fréchet differentiable function on U into a Banach space F then f is k-times directionally differentiable. Further f is a C^k-mapping iff the mappings $x \to T_x^i$ are continuous.

Before concluding the introductory chapter we recall two useful results from differential analysis of functions of a real variable.

1.3.5 Markov's inequality

Let P be a polynomial of degree n in a real variable. If $a < b$ are two real numbers, then

$$\sup_{a \leq t \leq b} |P'(t)| \leq \frac{n^2}{(b-a)} \sup_{a \leq t \leq b} |P(t)| .$$

1.3.6 Inequalities for the derivatives of a function

If f is k-times continuously differentiable real valued function on an open interval $]a,b[\subset R$ and if $\sup_{t \in]a,b[} |f(t)| = M_0$, $\sup_{t \in]a,b[} |f^k(t)| = M_k$, then

$$\sup_{t \in]a,b[} |f^{k-1}(t)| \leq \frac{8^{k-1}}{(b-a)^{k-1}} M_0 + \frac{(b+a)}{2} M_k .$$

For the inequalities in 1.3.5 and 1.3.6, see Todd [60], and Dieudonné [14].

Chapter 2

Smoothness Classification of Banach Spaces

In this chapter we discuss various differentiability properties of the norm in a Banach space. We are primarily interested in higher order differentiability of the norm.

2.1 Differentiability properties of norms

2.1.1 Definition

A Banach space $(E, \|\ \|)$ is said to be D^k-smooth (F^k-smooth) at a point $x \in E$, $x \neq 0$, if the $\|\ \|$ is k-times directionally differentiable at x (k-times Fréchet differentiable at x). E is D^k-smooth (F^k-smooth) if it is D^ksmooth at x (F^k-smooth at x) for all $x \neq 0$.

2.1.2 Remark

It follows from the definitions that if E is F^k-smooth then it is D^k-smooth.

2.1.3 Proposition

Let E be D^k-smooth at x, and (*) $\|x+ty\| = \|x\| + \sum_{i=1}^{k} t^i T_x^i(y^{(i)}) + \theta_x(ty)$, $y \in E$, be the expansion assured by the D^k-smoothness at x. Then

(1) E is D^k-smooth at λx, $\lambda \neq 0$, and $T_{\lambda x}^{\ i} = \dfrac{(\text{sign } \lambda)^i}{|\lambda|^{i-1}} T_x^i$,

(2) (i) $T_x^2(y,y) \geq 0$ for all $y \in E$, and

(ii) if T_x^2 is considered as a map on $E \to E^*$, range $T_x^2 \subset x^{\perp}$.

Proof

(1) follows at once by noting that $\|\lambda x+ty\| = |\lambda|\ \|x+ty\| =$
$= |\lambda|\{\|x\| + \sum_{i=1}^{k} (\frac{t}{\lambda})^i T_x^i(y^{(i)}) + \theta_x(\frac{ty}{\lambda})\}$, and for a fixed λ, $\lambda \neq 0$,
$\left|\dfrac{\theta\lambda_x(ty)}{t^k}\right| \to 0$ as $t \to 0$.

To prove (2) enough to verify that $T_x^2(y,y) \geq 0$, if $\|x\| = 1$, for all $y \in E$, by the preceding part of the proposition. Let $z \in E$, such that $T_x^1(z) = 0$. Then since $T_x^1(x) = ||x|| = 1$, as noted in 1.1.1, it is verified that (*) $||cx+tz|| \geq |c| \cdot ||x||$, for all real numbers c and t. Thus $||x+tz|| = ||x|| + t^2T_x^2(z,z) + \theta_x(tz) \geq ||x||$. Since $\frac{\theta_x(tz)}{t^2} \to 0$, as $t \to 0$, it follows that $T_x^2(z,z) \geq 0$. Also from the Taylor expansion of $||x+tx||$, it is verified that $T_x^2(x,x) = 0$. Now if $y \in E$, consider the unique decomposition of $y = \lambda x + z$, with $T_x^1(z) = 0$. Further $||x + \lambda y|| = ||x + \lambda tx + tz|| = ||x|| + tT_x^1(\lambda x + z) + t^2T_x^2(\lambda x + z, \lambda x + z) + \theta_x(t(\lambda x + z)) \geq |1 + \lambda t| = 1 + \lambda t$ for any fixed λ if $|t|$ is sufficiently small. Hence $t^2T_x^2(\lambda x + z, \lambda x + z) + + \theta_x(t(\lambda x + z)) \geq 0$. Dividing by t^2, and proceeding to the limit, it is verified that $T_x^2(y,y) = T_x^2(\lambda x + z, \lambda x + z) \geq 0$, verifying 2(i). To prove 2(ii) note that since $T_x^2(x,x) = 0$, the preceeding inequality implies $T_x^2(x,z) = 0$ for all z, such that $T_x^1(z) = 0$ i.e. $T_x^2(z) \perp x$. Now again considering the decomposition $y = \lambda x + z$, it is verified that $T_x^2(x,y) = 0$, completing the proof of 2(ii).

We proceed to discuss a fundamental isomorphism theorem which is very useful in various discussions in the classification to follow. If E is a Banach space, then a norm $|| \; ||_1$ on E is admissible if $(E, || \; ||_1)$ is isomorphic with E.

2.1.4 Theorem

If a Banach space E is D^2-smooth and its conjugate E^* is isomorphic with a D^2-smooth Banach space, then E is isomorphic with a Hilbert space.

Before proceeding to the proof we note if E is not isomorphic with a Hilbert space, and if T^2 is any symmetric continuous bilinear form on E then $\inf_{\substack{||y||=1 \\ y \in E_0}} |T^2(y,y)| = 0$ whenever E_0 is a closed subspace of finite codimension.

2.1.5 Lemma

If $(E, || \; ||)$ is D^2-smooth, and E^* is isomorphic with a

D^2-smooth space then there are admissible norms $\| \ \|_1$, $\| \ \|_2^1$ on E, E* respectively such that if

(i) $\|g\|_1^* \leq \|g\|_2^1$ for all $g \in E^*$,

(ii) $\exists f \in E^*$, $x \in E$, $\|x\| = 1$

such that $\|f\|_1^* = \|f\|^* = \|f\|_2^1 = f(x) = 1 = \|x\|_1$.

Proof

Let $(E, \| \ \|)$ be D^2-smooth and $\| \ \|_0^1$ be an admissible D^2-smooth norm on $(E^*, \| \ \|^*)$. Without loss of generality we may assume that $\|g\|^* \leq \|g\|_0^1 \leq A\|g\|^*$ for some constant A, for all $g \in E^*$. Choose $f \in E^*$, $x \in E$ such that $f(x) = 1 = \|f\|^* = \|x\|$. Now each $z \in E$ admits a decomposition $z = \lambda x + y$, where $f(y) = 0$. Define $\|z\|_1 = (\lambda^2 + \|y\|^2)^{1/2}$. It is verified that $\| \ \|_1$ is a D^2-smooth admissible norm on $(E, \| \ \|)$. Further $\|x\|_1 = 1$, and $\|f\|_1^* = 1$. Now consider the representation of each function $g \in E^*$, as $g = af + h$, where $e_x(h) = h(x) = 0$, e_x being the evaluation at x. Define $\| \ \|_2^1$ on E* by setting $\|g\|_2^1 = [a^2 + (\|g\|_0^1)^2]^{1/2}$; $\| \ \|_2^1$ is an admissible D^2-smooth norm on $(E^*, \| \ \|^*)$. If $g \in E^*$,

$$\|g\|_1^* = \sup_{\lambda^2 + \|y\|_1^2 \leq 1} g(\lambda x + y) \leq \sup_{\lambda^2 + \|y\|_1^2 \leq 1} \{|a| \, |\lambda| + \|h\|_1^* \|y\|_1\}$$

$$\leq (a^2 + \|h\|_1^{*2})^{1/2} = [a^2 + (\|h\|_0^1)^2]^{1/2}$$

since $\|y\|_1 = \|y\|$ for all $y \in f^{-1}(0)$. Clearly $\| \ \|_1^*$ is isomorphic with $\| \ \|_2^1$. It is also verified that

$$\|f\|_2^1 = \|f\|^* = \|f\|_1^* = f(x) = \|x\|_1 = \|x\| = 1.$$

Before proceeding to the proof of the theorem note that if T_x^1, T_x^2 (T_f^1, T_f^2) are the linear functional and bilinear form respectively on $E(E^*)$ such that,

$$\|x + ty\| = \|x\|_1 + tT_x^1(y) + t^2 T_x^2(y,y) + \theta_x(ty)$$

and

$$\|f + ty\|_2^1 = \|f\|_2^1 + tT_f^1(g) + t^2 T_f^2(g,g) + \theta_f(tg),$$

then from the uniqueness of support functions at $x(f)$ to the unit balls of $(E, \| \|_1)((E^*, \| \|_2^1)$ it follows that $T_x^1 = f$, and $T_f^1 = e_x$.

Proof of Theorem 2.1.4

Let $f, x, \| \|_1, \| \|_2^1$ be as in the lemma. We write each $g \in E^*$ as $g = af + h$, with $h(x) = 0$, a, of course, depending on g. Since the range $T_f^2 \subset f^\perp$ considering T_f^2 as a linear transformation on E^* into $(E^*, \| \|_2^1)^*$, it follows that $T_f(g,g) = T_f(h,h)$. Since T_f^2 is continuous, let $k > 0$ be such that (A) $T_f^2(g,g) \leq k \|g\|_1^{*2}$, for all $g \in E^*$. Now if y in E, h in E^* are such that $f(y) = 0$, and $h(x) = 0$, then

$$(f + th)(x + ty) = 1 + t^2h(y) \leq \|f + th\|_1^* \|x + ty\|_1$$

$$\leq \|f + th\|_2^1 \|x + ty\|_1 .$$

Now from the expansions $\|f + th\|_2^1$, $\|x + ty\|_1$ assured by the twice directional differentiability, it follows from our choice of h, x, since the first derivatives of $\| \|_2^1$, $\| \|_1$ at f and x respectively are e_x, and f, that

$$1 + t^2h(y) \leq 1 + t^2(T_f^2(h,h) + T_x^2(y,y)) + \psi(t,h,x,y,f)$$

where $\frac{\psi(t, \ldots)}{t^2} \to 0$ as $t \to 0$. Hence

(B) $h(y) \leq T_f(h,h) + T_x(y,y)$, for all $h \in E^*$, $y \in E$ with $h(x) = 0$, and $f(y) = 0$. Now if E is not isomorphic with a Hilbert space then $\inf\limits_{\substack{f(y)=0\\ \|y\|_1=1}} T_x^2(y,y) = 0$. Hence there exists $y_0 \in E$, $f(y_0) = 0$, $y_0 \neq 0$ such that $T_x^2(y_0,y_0) \leq \frac{1}{16k} \|y_0\|_1^2$ where k is as in (A). Let $g_0 \in E^*$ be such that (C) $g_0(y_0) = \|g_0\|_1^* \|y_0\|_1$. Without loss of generality we may assume that $\|g_0\|_1^* = \frac{\|y_0\|}{6k}$. Let $g_0 = \lambda f + h_0$ where $h_0(x) = 0$. As noted earlier, $T_f^2(h_0,h_0) = T_x^2(g_0,g_0)$. Hence from (B) it follows that

$$h_0(y_0) = \|g_0\|_1^* \; \|y_0\|_1^* = \frac{\|y_0\|_1^2}{6k}$$
$$\leq \frac{\|y_0\|_1^2}{36k} + \frac{1}{16k}\|y_0\|_1^2 ,$$

a contradiction completing the proof.

Before we pass on to the next topic we mention the following result, Meshkhov [42] , a stronger version of the preceeding theorem, when D^2 is replaced by C^2.

2.1.5(a) <u>Theorem</u>

If a Banach space E and its dual E^* admit nontrivial C^2-functions of bounded support, then E must be isomorphic with a Hilbert space.

2.1.6 <u>Definition</u>

A Banach space E is said to be C^k-smooth if the norm $\| \; \|$ of E is k-times continuously differentiable away from 0. It is UF^k-smooth if the norm is k-times uniformly Fréchet differentiable over the sphere $S_E = \{x \mid \|x\| = 1, \; x \in E\}$ of E. It is boundedly k-times Fréchet differentiable BF^k-smooth) if the norm is k-times Fréchet differentiable at $x \neq 0$, and the map $x \to T_x^k$, on $E \sim \{0\}$ into the Banach space of continuous i-linear forms is bounded over the sphere S_E i.e., $\sup_{x \in S_E} \|T_x^i\| < \infty$, $1 \leq i \leq k$.

2.1.7 <u>Theorem</u>

A Banach space $(E, \| \; \|)$ is U_F^k-smooth iff the norm $\| \; \|$ is uniformly k-times continuously differentiable in arbitrary ring shaped regions $R(\lambda, \mu) = \{x \mid \lambda < \|x\| < \mu\}$, for $0 < \lambda < \mu$. Every UF^k-smooth space is UF^i-smooth $1 \leq i \leq k$, as well as BF^k-smooth.

Before proceeding to the proof we note that $\| \; \|$ is uniformly continuously k-times differentiable in $R(a,b)$, $0 < a < b$ iff $\| \; \|$ is uniformly continuously k-times differentiable over the sphere S_E. This is an easy consequence of the fact that $\| \; \|$ is positively homogeneous, and $T^i_{\lambda x} = \frac{(\text{sign})^i}{|\lambda|^{i-1}} T^i_x$, iff $\lambda \neq 0$, $1 \leq i \leq k$, if the norm is k-times directionally differentiable, see 2.1.3.

Proof

(i) $UF^k \Rightarrow BF^i$, $1 \leq i \leq k$, for consider the expansion

(A) $\|x+ty\| = \|x\| + \sum_{i=1}^{k} t^i T^i_x(y^{(i)}) + \theta_x(ty)$ at an arbitrary $x \in S_E$;

since E is UF^k-smooth if $\epsilon > 0$ there exists a δ, $0 < \delta < 1$, δ independent of $x \in S_E$, $y \in S_E$, such that if $|t| \leq \delta$, $|\theta_x(ty)| \leq \epsilon |t|^k$. Thus from (A)

$$\left|\sum_{i=1}^{k} t^i T^i_x(y^{(i)})\right| \leq |t| + \epsilon|t|^k \leq 1 + \epsilon, \quad \text{for } |t| \leq \delta.$$

Applying Markov's inequality 1.3.5 to the successive derivatives of the polynomial $P(t) = \sum_{i=1}^{k} t^i T^i_x(y^{(i)})$, it follows that

$$|P^j(0)| = |j!\ T^j_x(y^{(i)})| \leq \frac{(k^2)(k-1)^2 \ldots (k-j+1)^2}{(\delta^2)^j}(1 + \epsilon),\ 1 \leq j \leq k.$$

Hence $\sup_{\|x\|=1} \|T^j_x\| < \infty$, $1 \leq j \leq k$, by 1.3.2(A). Hence E is BF^i-smooth, $1 \leq i \leq k$. Now since E is UF^k-smooth and BF^i-smooth it follows that E is UF^i-smooth $1 \leq i \leq k$. Let us denote $\sup_{\|x\|=1} \|T^j_x\|$ by M_j.

(ii) $UF^k \Rightarrow \|\ \ \|$ is uniformly k-times continuously differentiable in $R(a,b)$, a,b arbitrary positive numbers with $0 < a < b$. As already noted it suffices to show that the mappings $x \to T^i_x$ are uniformly continuous on S_E into the Banach spaces $B^i(E)$, for $1 \leq i \leq k$. Let $\{x_n\}$, $\{y_n\}$ be two sequences in S_E such that $\|x_n-y_n\| \to 0$. Applying expansion (A) to $\|x_n+ty\|$, $\|y_n+ty\|$, $\|y\| = 1$ it follows that for a given $\epsilon > 0$,

$$(*)\quad \left|\sum_{i=1}^{k} t^i (T^i_{x_n} - T^i_{y_n})(y^{(i)})\right| \leq \|x_n-y_n\| + |\theta_{x_n}(ty) - \theta_{y_n}(ty)|$$

$$\leq \|x_n-y_n\| + 2\ |t|^k, \quad \text{if } |t| \leq \delta(\epsilon),$$

where $\delta(\epsilon)$ is independent of n, and $y \in S_E$. From the preceding inequality it follows that

$$|(T^1_{x_n} - T^1_{y_n})y| \le \frac{\|x_n - y_n\|}{t} + 2\epsilon|t|^{k-1} + |\sum_2^k t^{i-1}(T^i_{x_n} - T^i_{y_n})(y^{(i)})|$$

$$\le \frac{\|x_n - y_n\|}{t} + 2\epsilon|t|^{k-1} + \sum_2^k 2M^i|t|^{i-1}$$

$$\le \frac{\|x_n - y_n\|}{t} + 3\epsilon \quad , \quad \text{if} \quad 0 < |t| \le \delta_1$$

where δ_1 depends only on ϵ, and $\{M^i\}_{i=2}^k$. Now fixing a value of t, $0 < |t| \le \delta_1$, and proceeding to the limit as $n \to \infty$, it follows that the map $x \to T^1_x$ is uniformly continuous on S_E. By induction, using an argument similar to the above, it follows that the maps $x \to T^i_x$, $1 \le i \le k$, are uniformly continuous on S_E.

(iii) We complete the proof of the theorem by showing that if the $\| \ \|$ is uniformly k-times continuously differentiable in $R(a,b)$ or equivalently in $R(1/2,3/2)$, then E is UF^k-smooth. Thus if $\epsilon > 0$ there is a $\delta(\epsilon) > 0$, which may be assumed to be $< 1/2$, such that if $\|x-y\| < \delta(\epsilon)$, $x, y \in R(1/2,3/2)$, then $\|T^k_x - T^k_y\| < \epsilon$. Now if $x \in S_E$ and if $\|x-y\| < \delta(\epsilon)$, then $[x,y] \subset R(1/2,3/2)$, and by Lagrange formula stated in 1.3 we have

$$\|y\| = \|x\| + \sum_{i=1}^{k-1} T^i_x((y-x)^{(i)}) + R_k, \quad \text{where}$$

$$R_k = \left(\int_0^1 \frac{(1-s)^{k-1}}{(k-1)!} (k!)\, T^k_{x+s(y-x)}\right) ds\,(y-x)^{(k)}.$$

Since $R_k = \int_0^1 (1-s)^{k-1} k(T^k_{x+s(y-x)} - T^k_x)\, ds\ (y-x)^{(k)} + T^k_x((y-x)^{(k)})$, it follows, from our choice of $\delta(\epsilon)$, if $x \in S_E$, and $\|h\| < \delta(\epsilon)$, then $\|x+h\| = \|x\| + \sum_{i=1}^{k} T^i_x(h^{(i)}) + \theta_x(h)$ where $|\theta_x(h)| \le \epsilon\|h\|^k$. Since the choice of $\delta(\epsilon)$ is clearly independent of $x \in S_E$, E is UF^k-smooth completing the proof of the theorem.

2.1.8 Remark

Since the only properties of the norm $\| \ \|$ which have been used are absolute homogenity, convexity and uniform continuity of the norm on bounded domains away from origin, Theorem 2.1.7 extends to the case when $\| \ \|$ is replaced by $\| \ \|^p$ where p is a positive number ≥ 1.

2.1.9 Theorem

If a Banach space is C^k-smooth and $\sup_{\|x\|=1} \|T_x^k\| < \infty$, then it is BF^i-smooth, $1 \leq i \leq k$.

Proof

The hypothesis implies that the $\| \ \|$ of E is k-times continuously differentiable in $R(1/2,3/2)$. Now let x,h be aribitrarily chosen from S_E. Then $[x - \frac{h}{2}, x + \frac{h}{2}] \subset R(1/2,3/2)$. Consider the function f defined on $]-1/2, 1/2[$ into R, with $f(t) = \|x+th\|$. Clearly f is a C^k-function, and since

$$\|x + th + \xi h\| = \|x + th\| + \sum_{i=1}^{k} \xi^i T^i_{x+th}(h^{(i)}) + \theta_{x+th}(\xi h)$$

where $\frac{\theta_{x+th}(\xi h)}{\xi k} \to 0$ as $\xi \to 0$, and $t \to T^i_{x+th}(h^{(i)})$ is continuous in $]-1/2, 1/2[$, it follows that the successive derivatives of f are given by $f^i(t) = i!\, T^i_{x+th}(h^{(i)})$, $1 \leq i \leq k$. Since the BF^k-smooth property of the norm over S_E implies that the norm is BF^k-smooth in $R(1/2,3/2)$ there exists a positive number M_k such that $|T^k_y(h^{(k)})| \leq k!M_k$, uniformly over $(y,h) \in R(1/2,3/2) \times S_E$. Thus $|f^k(t)| \leq k!M_k$, $-1/2 < t < 1/2$. Further $|f(t)| \leq S/2$ in $]-1/2,1/2[$. Now applying the inequality 1.3.6 it follows that

$$|(k-1)!\, T_x^{k-1}(h^{(k-1)})| = |f^{k-1}(0)|$$

$$\leq \sup_{|t| \leq \frac{1}{2}} |f^{k-1}(t)| \leq C,$$

for some constant C independent of $(x,h) \in S_E x S_E$. Thus $\| \ \|$ is BF^{k-1}-smooth. Repeating this argument it follows that $\| \ \|$ is BF^i-smooth, $1 \leq i \leq k$, as desired.

2.2 Differentiability of norms in classical Banach spaces

In this section we discuss the differentiability properties of norms in certain well known Banach spaces, Sundaresan [57]. In what follows if p is any positive number, we denote the integral part of p by $E(p)$.

(X,Σ,μ) is an arbitrary measure space with μ a positive measure. Before discussing the differentiability of the norm in $L_p(\mu)$, we state a useful lemma which is an easy consequence of Taylor's theorem. In what follows we assume $1 \leq p \leq \infty$, and μ is nontrivial so that $L_p(\mu)$ is of dimension ≥ 2.

2.2.1 Lemma

If a,b are any two real numbers there exists a constant M depending only on p, such that

$$\left| |a+b|^p - |a|^p - \sum_{i=1}^{E(p)} \binom{p}{i} |a|^{p-i} (\operatorname{sgn} a)^i b^i \right| \leq M|b|^p.$$

The lemma follows from the Taylor series expansion of $|a+b|^p$ with the Lagrange form of the remainder if $a \neq 0$.

2.2.2 Theorem

The norm in the Banach space $L_p(\mu)$ is (1) uniformly (p-1) times differentiable over bounded sets away from 0, if p is an odd integer, (2) uniformly k-times differentiable, over bounded sets away from 0, for all $k \geq 1$ if p is an even integer, and (3) uniformly $E(p)$-times differentiable, over bounded sets away from 0, if p is not an integer.

Proof

Let $p > 1$, and $f \in L_p(\mu)$, $f \neq 0$. Consider the functions T_f^i defined on $L_p(\mu) \to R$ by setting

$$T_f^i(g^{(i)}) = \binom{p}{i} \int_X |f|^{p-i} (\operatorname{sgn} f)^i g^i d\mu, \quad 1 \leq i < p.$$

Then

$$|T_f^i(g^{(i)})| \le \binom{p}{i}\left(\int_X |f|^p d\mu\right)^{1/p}\left(\int_X |g|^p d\mu\right)^{1/p},$$

by applying Hölder's inequality after noting that $|f|^{p-i} \in L_{p/i}(\mu)$ and $|g|^i \in L_{p/i}(\mu)$. Thus (2) $|T_f^i(g^{(i)})| \le \binom{p}{i} \|f\|_p \|g\|_p$. Further the function $T_f^i(g^{(i)})$ is the i-homogeneous polynomial associated with the symmetric i-linear form defined by

$$(*) \qquad T_f^i(g_1,g_2,\ldots,g_i) = \binom{p}{i}\int_X |f|^{p-i}(\text{sgn } f)^i g_1 g_2 \ldots g_i \, d\mu.$$

where P is the set of all permutations of (1,2,...,i). The continuity of the multilinear form (*) follows from the inequality (2) together with the remarks in 1.3.1 or by direct verification. Now from the lemma 2.2.1, it follows that if $f,g \in L_p(\mu)$, $\|f\|_p = 1$,

$$\left| \|f+g\|^p - \|f\|^p - \sum_{i=1}^{E(p)-1} \binom{p}{i}|f|^{p-i}(\text{sgn } f)^i g^i \right|$$

$$\le \binom{p}{E(p)}\int_X |f|^{p-i}(\text{sgn } f)^i g^i d\mu + M\|g\|_p^p$$

$$\le C\|g\|_p^p$$

where C does not depend on f or g. Hence the function $\| \;\|^p$ is uniformly (E(p)-1) times differentiable over the unit sphere of $L_p(\mu)$. Since $\|\lambda f\|^p = |\lambda|^p \|f\|^p$, $\| \;\|^p$ is uniformly i-times continuously differentiable in R(1/2,3/2) for $1 \le i \le E(p)-1$. Hence $\| \;\|$ has also the same property. Thus $\| \;\|$ is uniformly i-times differentiable over bounded sets away from 0, for $1 \le i \le E(p)-1$.

<u>Case 1</u>. Let p be an even integer. In this case inequality (1) holds with M = 0. Further the p-homogeneous form

$$T_f^p(g^{(p)}) = \int_X (g)^p d\mu$$

$$= \|g\|_p^p$$

is well defined and associated with bounded p-linear symmetric form.

We can introduce the i-homogeneous forms $T_f^i(g^{(i)}) \equiv 0$ for $i > p$, and for any integer $k \geq p$ we do have

$$\left| \|f+g\|^p - \|f\|^p - \sum_{i=1}^{k} T_f^i(g^{(i)}) \right| = 0.$$

Now proceeding as in the preceding paragraph it is verified that $\| \ \|$ is uniformly k-times differentiable over bounded sets away from 0, for all $k \geq 1$ in the case when p is even.

Case 2. Let p be an odd integer. Consider the map $f \to T_f^{p-1}$ on $L_p(\mu)$ into $B^{p-1}(L_p(\mu))$. This function is not even once directionally differentiable. For choose the disjoint measurable sets $A, B \in \Sigma$ of finite measure. Let $f = \chi_A$, $g = \chi_B$. It is verified that

$$\lim_{t\to 0} \frac{T_{f+tg}^{p-1} - T_f^{p-1}}{t} \quad \text{does not exist.}$$

Case 3. Let p be not an integer, so that $p \neq E(p)$. In this case considering the homogeneous form of degree $E(p) = k$, defined by

$$T_f^k(g^{(k)}) = \binom{p}{k} \int_X |f|^{p-k}(\operatorname{sgn} f)^k g^k d\mu ,$$

and arguing as in the Case 1 it is verified that the $\| \ \|$ is uniformly $E(p)$ times differentiable. Choosing f, g as in the Case 2 it is verified that the map $f \to T_f^{E(p)}$ is not even once directionally differentiable at f. This completes the proof of the theorem.

2.2.3 Remark

The norm in the Banach space $L_1(\mu)$ is not even once directionally differentiable away from 0. This is easily verified by choosing f, g as in the Case 2 in the preceding proof. Similarly the norm in $L_\infty(\mu)$ is not once directionally differentiable. For example, let f be the function $\chi_A - \chi_B$, where A, B are two disjoint measurable sets of finite measure. The norm is not directionally differentiable at such a function f.

We state the following theorem concerning the Banach spaces $C(X)$, where X is an infinite compact Hausdorff space due to Sundaresan [58], and Cox and Nadler [10] independently.

2.2.4 Proposition

The norm in $C(X)$ is directionally differentiable at a function $f \neq 0$ iff there is only one point $x \in X$, such that $|f(x)| = \|f\|$. It is Fréchet differentiable iff there is only one point $x \in X$ such that $|f(x)| = \|f\|$, and such a point x is isolated in X. Further if it is Fréchet differentiable at f, it is k-times Fréchet differentiable at f for all $k \geq 1$.

2.2.5 Remark

From Proposition 2.2.4 it follows that the norm in $C[0,1]$ is not Fréchet differentiable at any f, and is not directionally differentiable away from 0.

Since we are interested in differentiability properties modulo isomorphism we recall two important theorems in this direction. The first one has been observed by Restrepo [53], and is a consequence of Bishop-Phelps density lemma [8] and a renormability theorem of Klee [33]. The second theorem is due to Kuiper, and the proof appears in Bonic and Frampton [9]. In this connection, see also Yamamuro [67].

2.2.6 Theorem [Restrepo]

A separable Banach space E admits an equivalent Fréchet differentiable norm iff the dual of E is separable.

Proof

Let E be a separable Banach space, and $\| \; \|$ be a Fréchet differentiable norm on E, equivalent to the original norm. Then the map $G: E \sim \{0\} \to E^*$ is continuous where $G(x)$ is the derivative of the $\| \; \|$ at x. Thus the set $G(E \sim \{0\})$ contains all the support functionals w.r.t. the norm, $\| \; \|$. Hence $G(E \sim \{0\})$ is dense in

E^* by Bishop-Phelps density lemma [8]. Since G is continuous and E is separable it follows that E^* is separable.

Conversely let E^* be separable. Then by Klee's theorem there is a norm $\|\ \ \|$ on E equivalent to the original norm such that the $\|\ \ \|$ has the property (α): if $x_n \to x$ weakly, and $\|x_n\| = \|x\|$ for all $n \geq 1$, then $x_n \to x$ in the norm topology of E. Since the property α of a norm implies it is of class C^1, see [11], the proof is complete.

2.2.7 Theorem (Kuiper)

The Banach space c_0 admits an equivalent norm which is C^∞-away from origin.

Proof

Let $\alpha_1 : R \to R$ be the function defined by $\alpha_1(t) = \exp(-t^{-2})$ if $t > 0$, and $\alpha_1(t) = 0$ if $t \leq 0$. Let $\alpha_2(t) = \alpha_1(t-1)\alpha_1(2-t)$, and $\alpha_3(x) = \int_0^x \alpha_2(t)\,dt \Big| \int_0^2 \alpha_2(t)\,dt$. Thus α_1, α_2, α_3 are C^∞-functions defined on $R \to R$. The support of α_2 is $[1,2]$, and the support of α_3 is $[1,\infty[$. Further $\alpha_3(x) = 1$ if $x \geq 2$, $\alpha_3(x) = 0$ if $x \leq 1$, and $0 \leq \alpha_3(x) \leq 1$, if $1 \leq x \leq 2$. Define $h : R \to R$ by

$$h(x) = 1 - \alpha_3(|x|)$$

and $f(x_1, x_2, \ldots) = \prod_{i \geq 1} h(x_i)$, for $(x_1, x_2, \ldots)$ in c_0.

By our choice of $h(x)$, f is a real valued function defined on c_0, and since f reduces to a product of finitely many C^∞-functions locally, f is of class C^∞. Let $M = \{x \mid x \in c_0, \text{ and } f(x) \geq \frac{1}{2}\}$. M is verified to be a bounded convex symmetric subset of c_0, and contains as a subset the unit ball of c_0. Hence the Minkowski functional of M defines a norm N equivalent to the norm of c_0, and since N depends locally on finite number of coordinates in a C^∞ way it follows that N is of class C^∞ on c_0.

In recent years the vector valued Banach spaces of functions, known in the literature as Bochner function spaces, have been very extensively studied as they are found to be important in the study of vector valued random variables as well as in the study of operators on a L_p space into a Banach space.

2.2.8 Definition

If (T,Σ,μ) is a measure space and $(E,\|\ \|)$ is a Banach space the function spaces $L_p(E,\mu)$, are defined as follows:

$$L_p(E,\mu) = \{f \mid f:T \to E \text{ is measurable and } \int_T \|f(t)\|^p d\mu(t) < \infty\},$$

if $1 \leq p < \infty$

and

$$L_\infty(E,\mu) = \{f \mid f:T \to E \text{ is measurable and ess sup } \|f(t)\| < \infty\}.$$

(As usual, we are identifying functions f which agree μ a.e.) If T is the set of positive integers, and μ is the counting measure, $L_p(E,\mu)$ is usually denoted by $\ell_p(E)$. The vector spaces $L_p(E,\mu)$ are equipped with the norms

$$\|f\|_p = \left(\int_T \|f(t)\|^p d\mu(t)\right)^{1/p}, \text{ if } 1 \leq p < \infty, \text{ and}$$

$$\|f\|_\infty = \operatorname*{ess\,sup}_{t \in T} \|f(t)\|.$$

It is known that the spaces $L_p(E,\mu)$ with above norms are Banach spaces, and are generally referred to as Bochner spaces. For an account of these spaces see Dinculeanu [17], Dunford and Schwartz [19], Diestel and Uhl [13], and Hille and Phillips [26].

The differentiability of the norm in the spaces $L_p(E,\mu)$ are completely characterized in Leonard and Sundaresan [39]. We state these results here without proof, and refer the interested reader to the paper [39] for proofs.

2.2.9 Theorem [Leonard and Sundaresan]

Let $(E,\|\ \|)$ be a Banach space, (T,Σ,μ) be a measure space, and $1 < p < \infty$, then:

(i) The norm in $L_p(E,\mu)$ is differentiable away from zero if and only if the norm in E is differentiable away from zero.

(ii) If $p = 2$, the norm in $L_p(E,\mu)$ is twice continuously differentiable away from 0 iff E is a Hilbert space.

(iii) If p is an even integer, $p > 2$, the norm in $L_p(E,\mu)$ is p times continuously differentiable away from 0 if and only if $\|\cdot\|^p$, is a continuous homogeneous polynomial of degree p.

(iv) If p is an odd integer, the norm in $L_p(E,\mu)$ is $(p - 1)$ times continuously differentiable away from 0, and the $(p - 1)$th derivative of the norm in E is uniformly bounded on the unit sphere in E.

(v) If p is not an integer, and $E(p)$ is the integral part of p, the norm in $L_p(E,\mu)$ is $E(p)$-times continuously differentiable away from 0 iff the norm in E is $E(p)$-times continuously differentiable away from 0, and the $E(p)$th derivative of the norm, $\| \ \|$, is uniformly bounded on the unit sphere in E.

The differentiability of the norm in the Banach spaces $C(X,B)$, the vector-valued analogues of the classical spaces $C(X)$, is discussed in [10] and [58].

Let X be a compact Hausdorff space and $(B, \| \ \|)$ be a Banach space. Let $C(X,B)$ be the Banach space of continuous B-valued functions $f:X \to B$, with the norm $\|f\| = \sup\{\|f(t)\| \mid t \in X\}$.

2.2.10 <u>Theorem</u>

The norm in $C(X,B)$ is p-times (infinitely) continuously differentiable in a neighborhood of f, $f \neq 0$, iff there exists only one point $p \in X$ such that $\|f(p)\| = \|f\|$, and such a point p is isolated in X, and in such a case the norm in B is p-times (infinitely) continuously differentiable in a neighborhood of $f(p)$ in B

For a proof, see either [10] or [58].

We conclude this section stating the results on the smoothness of the Banach spaces of trace classes S_p.

Let H be a real Hilbert space, and $K(H)$ be the Banach space of compact operators on H into H with the usual supremum norm. If $A \in K(H)$, let A^* denote the adjoint of A. The S-numbers of the operator A, denoted by $S_n(A) = \lambda_n$, $n = 1,2,\dots$ where $\{\lambda_n\}$ is the non-increasing sequence of nonzero eigenvalues of the operator $(A^*A)^{1/2}$, each repeated the number of times equal to its multiplicity.

Let $1 \leq p < \infty$. We put

$$S_p = \{A \mid A \in K(H),\ \|A\|_p = (\sum_{n\geq 1} S_n^p(A))^{1/p} < \infty\}.$$

It is known that S_p is a Banach space under the norm $\|A\|_p$, and

$$\|A\|_p = (tr(A^*A)^{p/2})^{1/p}.$$

The differentiability properties of the norms in S_p, $1 \leq p < \infty$ are similar to the properties of the norms in $L_p(\mu)$, for the corresponding values of p. For these results we refer to Tomczak-Jaegerman [61].

2.3 UF^k-smooth spaces and ultrapowers

In this section we present certain results concerning finite representation and UF^k-smooth Banach spaces, and proceed to present a classification of superreflexive spaces. We start with a theorem concerning the duals of ultrapowers of superreflexive spaces.

Let T be an infinite set and U a nontrivial ultrafilter on T. We assume further that there is a positive function $\delta : T \to R$ such that $\lim_U \delta = 0$. For instance if T is the set of natural numbers then $\lim_U \delta = 0$ if $\delta(n) = 1/n$, and U is the ultra filter generated by the tails on T.

First we establish a lemma on ultrapowers $E(T,U)$ assuming (T,U) admits a δ-function described above.

2.3.1 Lemma

For a Banach space E, there exists an isometry $\sigma: E^*(T,U)$ into $(E(T,U))^*$, where $\sigma(\tilde{f})(\tilde{x}) = \lim_U f(t)(x(t))$ if $\tilde{f}$, $\tilde{x}$ are in $E^*(T,U)$, and $E(T,U)$ respectively, and $\{f(t)\}$, $\{x(t)\}$ are arbitrary functions in $\tilde{f}$ and $\tilde{x}$ respectively.

Proof

Noting that if $\tilde{y}$ is any member of an arbitrary ultrapower of E, and $\{y_1(t)\}$, $\{y_2(t)\}$ are two functions in $\tilde{y}$, then by the definition of the norm in the ultrapower it follows that $\lim_U \|y_1(t) - y_2(t)\| = 0$. Using this fact, together with the fact that $\{f(t)\}$, $\{x(t)\}$ are bounded E^* and E-valued functions respectively, it follows that $\sigma(\tilde{f})$ is well defined (independent of the representatives $\{f(t)\}$, $\{x(t)\}$) linear functional on $E(T,U)$. Further denoting the norms of $E(T,U)$, and $E^*(T,U)$ respectively by $\| \ \|_1$ and $\| \ \|_2$, we obtain

$$(1) \qquad |\sigma(\tilde{f})(\tilde{x})| \le \|\tilde{f}\|_2 \ \|\tilde{x}\|_1 .$$

Hence $\sigma(\tilde{f}) \in (E(T,U))^*$. σ is an isometry for if $\tilde{f} \in E^*(T,U)$, choose $x(t) \in E$ $\|x(t)\| = 1$ such that $f(t)(x(t)) \ge \|f(t)\|^* - \delta(t)$ where δ is the function described in 2.3., and $\{f(t)\}$ is an arbitrary member of $\tilde{f}$, and the norm of E is $\| \ \|$. Now $\sigma(\tilde{f})(\tilde{x}) = \lim_U f(t)(x(t)) \ge \lim (\|f(t)\| - \delta(t)) =$ $= \lim \|f(t)\| = \|\tilde{f}\|_2$. Thus $\|\sigma(\tilde{f})\|_1^* = \|\tilde{f}\|_2$.

2.3.2 Theorem

Assuming that the pair (T,U) admits a δ-function described above, $\sigma(E^*(T,U)) = (E(T,U))^*$ iff E is superreflexive, where the map σ is as described in the preceding lemma.

Proof

Let $\tilde{x} \in E(T,\Gamma)$. If $\{x(t)\} \in \tilde{x}$, choose $f(t) \in E^*$ such that $f(t)(x(t)) = \|x(t)\|$, $\|f(t)\|^* = 1$. Thus $\sigma(\tilde{f})(\tilde{x}) = \|\tilde{x}\|_1$. Hence $\sigma(E^*(T,U))$ is a norm determining subspace for $E(T,U)$. Also by the lemma, $\sigma: E^*(T,U) \to (E(T,U))^*$ is an isometry. It follows that

$$\|\tilde{f}\|_2 = \|\sigma(\tilde{f})\|_1^* = \sup_{\|\tilde{x}\|_1=1} |(\tilde{f})(\tilde{x})|.$$

Thus $\sigma(E^*(T,U))$ is strictly norming subspace for $E(T,U)$. Now if $\sigma(E^*(T,U)) = (E(T,U))^*$ then since each $\sigma(\tilde{f})$ attains its norm, as shown in Lemma 2.3.1, it follows from a theorem of R.C.James that $E(T,U)$ is reflexive. Hence E is superreflexive, see 1.2.3. Conversely if E is superreflexive then $E(T,U)$ is reflexive since $E(T,U) \ll E$. Thus the strictly norming subspace $\sigma(E^*(T,U))$ is all of $(E(T,U))^*$, completing the proof of the theorem.

2.3.3 Lemma

If E and F are isomorphic Banach spaces, then ultrapowers $E(T,U)$ and $F(T,U)$ are isomorphic.

Proof

Let $T: E \to F$ be an isomorphism on E onto F. If $\tilde{f} \in E(T,U)$ and $\{f(t)\} \in \tilde{f}$, define $\tilde{T}\tilde{f}$ to be the member of $F(T,U)$ determined by $\{Tf(t)\}$. Since T is uniformly continuous, it is verified that $\tilde{T}\tilde{f}$ is independent of the choice of $\{f(t)\} \in \tilde{f}$. Since T is bounded and linear and onto F, it follows that $\tilde{T}$ is a continuous linear operator onto $F(T,U)$. Since T is 1-1 it is verified from the definition of the norms in the ultrapowers, that $\tilde{T}$ is 1-1. Hence $\tilde{T}$ is an isomorphism. We refer to $\tilde{T}$ as the canonical extension of T.

2.3.4 Proposition

If a Banach space E is uniformly F^1-smooth then every ultrapower $E(T,U)$ is F^1-smooth. Further, if there is a function δ on T to positive real numbers, such that $\lim_U \delta(t) = 0$, and if $E(T,U)$ is F^1-smooth then E is uniformly F^1-smooth.

Proof

Before proceeding to the proof we note the following. Let E be F^1 - smooth, and T_x^1 is the first derivative of the norm of

E at $z \neq 0$. Let $0 \neq \tilde{x} \in E(T,U)$, and let $\{x(t)\} \in \tilde{x}$. If $\tilde{z} \in E(T,U)$, and $\{z(t)\} \in \tilde{z}$. Define $f(t) = 0$ if $x(t) = 0$, and $f(t) = T^1_{x(t)}(x(t))$ if $x(t) \neq 0$. It follows that $\lim_U f(t)$ is independent of the representatives $\{x(t)\}$, $\{z(t)\}$ of $\tilde{x}$, and $\tilde{z}$. We may assume that $\|\tilde{x}\| = 1$, and $\|z(t)\| \leq M$ for some $M \geq 0$, for all $t \in T$. Then $|f(t)| \leq M$ for all $t \in T$, so that $\lim_U f(t)$ exists. Now suppose $\{z_1(t)\} \in \tilde{z}$, and $\epsilon > 0$. Then there exists a set J_ϵ in U, such that for all $t \in J_\epsilon$ $\|z(t) - z_1(t)\| < \epsilon$, and $1/2 < \|x(t)\| < 3/2$. Hence $|T^1_{x(t)}(z(t) - z_1(t))| < \epsilon$ for $t \in J_\epsilon$. Since ϵ is arbitrary, the preceding implies that $\lim_U f(t)$ is independent of $\{z(t)\} \in \tilde{z}$.

Since the norm $\|\ \|$ is uniformly F^1-smooth, the map $x \to T^1_x$ is uniformly continuous in $R(1/2, 3/2) \subset E$ into E^*, repeating the preceding argument for distinct choices $\{x(t)\}$, $\{x_1(t)\}$ of $\tilde{x}$, it is verified that the $\lim_U f(t)$ is independent of the representative $\{x(t)\}$ of $\tilde{x}$. Now combining the preceding observations it follows that $\lim_U f(t)$ is well defined. Let us denote this limit by $\ell_{\tilde{x}}(\tilde{z})$. It is verified that $\ell_{\tilde{x}}$ is a linear functional over $E(T,U)$ and $\|\ell_{\tilde{x}}\| = 1$. Thus $\ell_{\tilde{x}} \in (E(T,U))^*$.

Now to complete the proof of the first part of the theorem note that for a given $\epsilon > 0$, there is a $\delta > 0$ independent of x, $x \in R(1/2, 3/2) \subset E$, such that

(B) $$\|x + h\| = \|x\| + T^1_x(h) + \theta_x(h)$$

where $|\theta_x(h)| \leq \epsilon\|h\|$ if $\|h\| < \delta$. Now let $\tilde{x}$, $\tilde{y}$, be in $E(T,U)$, each of unit norm, and $\{x(t)\}$, $\{y(t)\}$ be representative of $\tilde{x}$, $\tilde{y}$, respectively. We may assume that $\frac{1}{2} < \|x(t)\|$, $\|y(t)\| < \frac{3}{2}$. Hence if $\ell_{\tilde{x}}$ defined in the preceding paragraph is chosen for T^1_x, then

$$\|\tilde{x} + \tilde{y}\| = \|\tilde{x}\| + \lambda T^1_{\tilde{x}}(y) + \lim_U \theta_{x(t)}(\theta y(t)).$$

It follows from (B) that there is a $\delta_1 > 0$, such that

$|\theta_{x(t)}(\lambda(y(t)))| \leq \epsilon|\lambda|$ if $|\lambda| < \delta_1$, δ_1 independent of t in T. Hence $|\lim_U \theta_{x(t)}(\lambda y(t))| \leq \epsilon|\lambda|$ if $|\lambda| < \delta_1$, verifying that the $\| \ \|$ in $E(T,U)$ is F^1-smooth.

To prove the second part of the theorem let (T,U) admit a function $\delta > 0$ such that $\lim_U \delta(t) = 0$, and let $E(T,U)$ be F^1-smooth. If E is not uniformly F^1-smooth there exists a positive ϵ, two E-valued functions $\{x(t)\}$, $\{y(t)\}$ such that $\|x(t)\| = 1 = \|y(t)\|$, $\|x(t) - y(t)\| \leq \delta(t)$, but

(C) $$\|T^1_{x(t)} - T^1_{y(t)}\| \geq \epsilon \quad \text{for all } t \in T.$$

Since $\lim_U \delta(t) = 0$, it follows that if $\{x(t)\}$, $\{y(t)\}$ determine the elements $\tilde{x}$, $\tilde{y}$ in $E(T,U)$, then $\tilde{x} = \tilde{y}$, and $T^1_{\tilde{x}} = T^1_{\tilde{y}}$. However (C) guarantees the existence of $z(t) \in E$ $\|z(t)\| = 1$, for $t \in T$, such that $(T^1_{x(t)} - T^1_{y(t)})(z(t)) \geq \epsilon$. Arguing as in the preceding part, it is seen, if $\{z(t)\} \in \tilde{z}$, that $\lim_U T^1_{x(t)}(z(t)) = T^1_{\tilde{x}}(\tilde{z})$, $\lim_U T^1_{y(t)}(z(t)) = T^1_{\tilde{y}}(\tilde{z})$, and the preceding inequality implies $T^1_{\tilde{x}}(\tilde{z}) \neq T^1_{\tilde{y}}(\tilde{z})$ contradicting that $T^1_{\tilde{x}} = T^1_{\tilde{y}}$, completing the proof of the proposition.

2.3.5 <u>Theorem</u>

If F is a Banach space, then F is uniformly F^1-smooth, if and only if every Banach space $E<<F$ is F^1-smooth.

<u>Proof</u>

Suppose that $E<<F$ implies that E is F^1-smooth. Thus if N is the set of positive integers and U is a free ultrafilter on N, then $F(N,U)<<F$. Thus $F(N,U)$ is F^1-smooth. Thus from the prece proposition, since (N,U) admits a function of type δ (e.g. let $\delta(n) = 1/n$), F is uniformly F^1-smooth. Next suppose that F is uniformly F^1-smooth, and $E<<F$. Since E is isometric with a subspace of an ultrapower of F, E is F^1-smooth as seen by applying the

preceding proposition.

The following remark is crucial for the proposed classification of superreflexive spaces discussed in the concluding part of this section.

2.3.6 Remark

Since the relation $<<$ is transitive it follows from the preceding theorem that a Banach space F is uniformly F^1-smooth iff $E<<F$ implies that E is a uniformly F^1-smooth space.

Before extending the preceding theorem we recall a part of the theorem as a lemma here.

2.3.7 Lemma

If E is uniformly F^k-smooth then it is boundedly F^k-smooth.

We now proceed to extend the preceding theorem for uniformly F^k-smooth spaces, $k \geq 2$.

2.3.8 Theorem

If E is a uniformly F^k-smooth space, $k \geq 2$, then every ultrapower $E(T,U)$ of E is also uniformly F^k-smooth.

Proof

Let $\tilde{f}$, $\tilde{g} \in E(T,U)$ with $\|\tilde{f}\| = 1 = \|\tilde{g}\|$. We can assume that there are representatives $\{f(t)\}$, $\{g(t)\}$ members of $\tilde{f}$, $\tilde{g}$ respectively such that $\frac{1}{2} < \|f(t)\| < \frac{3}{2}$, $\frac{1}{2} < \|g(t)\| < \frac{3}{2}$. Since the norm in E is uniformly F^k-differentiable, it follows that if $\epsilon > 0$, then

$$\|f(t) + \xi g(t)\| = \|f(t)\| + \sum_{i=1}^{k} \xi^i T^i_{f(t)}(g^i(t)) + \theta_{f(t)}(\xi g(t))$$

where (*) $|\theta_{f(t)}(\xi g(t))| < \epsilon|\xi|^k$ if $|\xi| < \delta$, δ independent of $t \in T$. We also note that δ depends only on the ring $R(\frac{1}{2}, \frac{3}{2})$ of E, uniformly. Further since UF^k-smoothness implies BF^k-smoothness, it follows that (a) $L^i_{\tilde{f}}(g) = \lim\limits_{U} T^i_{f(t)}(g^i(t))$ exists. Further the limit $L^i_{\tilde{f}}(\tilde{g})$ is independent of the choice of the representatives

$\{f(t)\}$, $\{g(t)\}$ from $\tilde{f}$, $\tilde{g}$. This is verified by noting that (i) any homogeneous polynomial on a Banach space E is uniformly continuous, (ii) the map $x \to T_x^i$ is uniformly continuous on bounded domains of E, away from 0, into $B^i(E)$. The first observation assumes the independence of the limit in (a) from the choice of $\{g(t)\} \in \tilde{g}$, while the second observation implies that the limit in (a) is independent of the choice of $\{f(t)\} \in \tilde{f}$. Thus $L^i_{\tilde{f}}(\tilde{g}^{(i)})$ defines a continuous i-homogeneous polynomial on $E(T,U)$, $1 \leq i \leq k$. Thus

$$\lim_U \|f(t) + \xi g(t)\| = \lim_U \|f(t)\| + \sum_{i=1}^{k} L^i_{\tilde{f}}(\tilde{g}^{(i)}) + \lim_U \theta_{f(t)}(\xi g(t)).$$

Hence (b) $\|\tilde{f} + \xi\tilde{g}\| = \|\tilde{f}\| + \sum_{i=1}^{k} L^i_{\tilde{f}}(\tilde{g}^{(i)}) + \lim_U \theta_{f(t)}(\xi g(t))$.

Now from (*) it follows that

(c) $|\lim_U \theta_{f(t)}(\xi g(t))| \leq \epsilon|\xi|^k$ if $|\xi| < \delta$, since δ is independent of t. Further since the same δ serves uniformly in $R(\frac{1}{2}, \frac{3}{2})$ i.e., $|\theta_x(\xi y)| < \epsilon|\xi|^k$, if $x, y \in R(\frac{1}{2}, \frac{3}{2})$ and $|\xi| < \delta$, (b) and (c) imply that the norm in $E(T,U)$ is uniformly F^k-differentiable on bounded domains away from 0. Hence $E(T,U)$ is UF^k-smooth.

2.3.9 <u>Theorem</u>

If $E(T,U)$ is UF^k-smooth, then E is UF^k-smooth.

<u>Proof</u>

This follows at once by noting that E is isometric with a subspace of $E(T,U)$.

The preceding two theorems imply the following remark.

2.3.10 <u>Remark</u>

A Banach space E is UF^k-smooth iff every Banach space $F \ll E$ is UF^k-smooth.

It must be observed that the preceding result need not hold, if the uniformity hypothesis is dropped. We illustrate this with the following example.

2.3.11 Example

Consider the isomorph $(B, \| \|)$ of the classical Banach space c_0 such that B is C^k-smooth for all $k \geq 1$, assured by the theorem 2.2.7. It is well known [56] that every Banach space, in particular the classical Banach space $C[0,1]$, is finitely represented in c_0. Thus there is an isomorph of $C[0,1]$, say E, such that $E \ll B$. However $C[0,1]$ is not isomorphic with any C^1-smooth space, see theorem 2.2.6. This shows that the preceding remark fails in the absence of uniformity of smoothness (even when higher derivatives are available).

We now present a few corollaries of the Remark 2.3.10.

2.3.12 Corollary

If F_1, F_2 are two uniformly F^2-smooth Banach spaces and E is a Banach space such that $E \ll F_1$, $E^* \ll F_2$, then E is isomorphic with a Hilbert space. In particular if $E \ll \ell_{p_1}$, $E^* \ll \ell_{p_2}$, $2 \leq p_1, p_2 \leq \infty$, then E is isomorphic with a Hilbert space.

Proof

The hypothesis in the corollary together with Remark 2.3.10 imply that E, and E^* are F^2-smooth. Hence it follows from Theorem 2.1.4 that E is isomorphic with a Hilbert space.

Next we state a property of Banach spaces E which are finitely representable in $\ell_p(L_p)$, p an even integer. It is shown in Theorem 2.2.2 that these spaces are UF^k-smooth for all $k \geq 1$, and that the norms in these spaces are determined by homogeneous polynomials of degree p, in the sense that $\|x\|^p = \emptyset(x)$, where $\emptyset(x)$ is a homogeneous polynomial of degree p. Now if $F = \ell_p(L_p)$, p even, then the norm in any ultrapower $F(T,U)$ is determined by a homogeneous polynomial of degree p, as one can verify from the definition of the norm in $F(T,U)$. Thus we arrive at the following.

2.3.13 Remark

If $E<<F$ where $F = \ell_p(L_p)$, p an even integer, then the norm in E is determined by a homogeneous polynomial of degree p.

2.3.14 Corollary

If E is an infinite dimensional Banach space which is uniformly F^k-smooth but not uniformly F^{k+1}-smooth then none of the spaces ℓ_p, $1 \leq p \leq k$, p not an even integer is finitely representable in E.

This is a consequence of Theorem 2.1.2 and Remark 2.1.14.

Let $L_p(E,\mu)$ be the Bochner spaces associated with a measure space (X,Σ,μ), and a Banach space E, see 2.2.8.

2.3.15 Corollary

A necessary condition for $L_p(E,\mu)$ to be finitely representable in a Banach space F, where $F = \ell_m(L_m(\lambda))$, is that $L_p(\mu)$ and E must be at least of the same uniform smoothness class as ℓ_m. Thus for instance, $L_2(\ell_3,\mu)$ is not finitely representable either in $\ell_2(L_2)$ or in ℓ_5.

2.4 Classification of superreflexive spaces

We now proceed to develop a method to classify superreflexive spaces in terms of finite representability and smoothness. As already noted in Chapter 1, a Banach space is superreflexive if and only if it is isomorphic with a uniformly F^1-smooth Banach space. Motivated by this, and our results in Section 2.1 we define that a Banach space E is k-superreflexive if it is isomorphic with a Banach space which is uniformly F^k-smooth. If E is k-superreflexive for all $k \geq 1$, then E is said to be ω-superreflexive. Note that 1-superreflexive spaces are superreflexive spaces as defined in Chapter 1.

2.4.1 Proposition

A Banach space E is k-superreflexive iff $F<<E$ implies F is k-superreflexive.

<u>Proof</u>

Let E be k-superreflexive, and F<<E. Thus there is a uniformly F^k-smooth Banach space E_1 isomorphic with E. Since F is isometric with a subspace of an ultrapower E(T,U) of E, and since E(T,U) is isomorphic with $E_1(T,U)$ canonically, F is isomorphic with a subspace of $E_1(T,U)$. By Remark 2.3.6 and Theorem 2.3.8, it follows that $E_1(T,U)$ is UF^k-smooth. Hence F is k-superreflexive. The converse is immediate.

From the smoothness properties of $\ell_p(L_p(\mu))$ spaces, where the measure μ is not supported by finitely many atoms, it follows that there are Banach spaces which are k-superreflexive but which are not k^1-superreflexive for any $k^1 > k$. Finally it is of interest to note that if E is k-superreflexive, $k \geq 2$, then E* is 1-superreflexive, and not m-superreflexive for $m \geq 2$, unless E is isomorphic with a Hilbert space in which case E, and E* are ω-superreflexive.

Chapter 3

Smooth Partitions of Unity on Banach Spaces

3.1 S-categories, Smooth pairs, and S-partitions of Unity

The main problem discussed in this chapter may be simply described as follows. Suppose E is a Banach space, and f is a nontrivial real valued function with bounded support on E satisfying a smoothness condition α. Possible choices for α include (i) $f \in D^n$ i.e. f is n-times differentiable, (ii) $f \in C^n$, (iii) $f \in C^n$ and the successive derivatives of f satisfy Hölder condition δ, $0 < \delta < 1$. In many situations in non-linear analysis on Banach spaces E it is of importance to know whether every open covering of E admits a partition of unity (locally finite) such that the functions involved in the partition satisfy a smoothness condition α. The answer is affirmative for many choices of α, and in all these cases the procedure to construct the partition of unity is the same. In order to develop a unified theory for these choices of α, Bonic and Frampton [9] developed the concept of S-categories. We do not need here the full generality in which it is discussed in [9]; however, we mention it for completeness, because it is useful in several situations in non-linear analysis.

3.1.3 Definition

A category S is a S-category if it has the following properties.

(A) The objects of S, denoted by $\mathcal{G}$, are open subsets of Banach spaces.

(B) If $U, V \in \mathcal{G}$, the set of morphisms $S(U,V)$ has the following properties:

(i) $f \in S(U,V) \Rightarrow f$ is a continuous function on $U \to V$.

(ii) $C^\infty(U,V) \subset S(U,V)$.

(iii) $f \in S(U,V)$, and $f(U) \subset W \in \mathcal{G} \Rightarrow f \in S(U,W)$.

(iv) If for any $x \in U$, there is a $W \in \mathcal{G}$ such that $x \in W \subset U$, and $f|W \in S(W,V)$, then $f \in S(U,V)$.

(v) If $f_i \in S(U_i,V_i)$, then $f_1 \times f_2 \in S(U_1 \times U_2, V_1 \times V_2)$.

The maps $f \in S(U,V)$ are referred to as S-smooth maps.

Note that if $f \in S$ and g is a C^∞-function, and if $g \circ f$ is defined, then $g \circ f \in S$. Further if $f_i \in S(U,V)$, $1 \le i \le n$, $g \in C^\infty(V^{(n)},V)$ where $V^{(n)} = \underbrace{V \times V \times \ldots \times V}_{n}$, then $g \circ (f_1 \times f_2 \times \ldots \times f_n) \circ d \in S(U,V)$, where d is the diagonal map on $U \to U^{(n)}$. Hence it follows that if V is a vector space (Banach algebra) then $S(U,V)$ is a vector space (Banach algebra), since the algebraic operations are C^∞-maps.

3.1.2 Examples

The following examples indicate morphisms inducing S-categories.

1. $D^n(U,V) = \{f | f:U \to V,\ f \text{ is } n \text{ times differentiable}\}$.
2. $C^n(U,V) = \{f | f:U \to V,\ f \text{ is } n \text{ times continuously differentiable}\}$.
3. $D^n_\alpha(U,V) = \{f | f \in D^n(U,V), \text{ and } \|D^n f(x+y) - D^n f(x)\| = O(\|y\|^\alpha)\}$.
4. $C^n_\alpha(U,V) = \{f | f \in C^n(U,V), \text{ and } \|D^n f(x+y) - D^n f(x)\| = o(\|y\|^\alpha)\}$.

In the context of the topics that interest us here, a special case of S-categories is to be emphasized.

3.1.2(a) Definition

A pair (M,S) is called a smooth pair if it satisfies the following properties:

(I) M is a metric space, and S is a set of continuous functions on $M \to R$.

(II) If $f:M \to R$ is a continuous function, and if there is an open covering $\mathcal{G}$ of M such that there is a set $\{g_U | U \in \mathcal{G}\} \subset S$ with $g_U|U = f|U$, then $f \in S$.

(III) If $g_i \in S$, $1 \le i \le n$, and $\phi \in C^\infty(R^n,R)$, then $\phi \circ (g_1,g_2,\ldots,g_n) \in S$.

3.1.3 Remark

Let, for a set S of continuous functions on $M \to R$, $\mathcal{U}_S = \{f^{-1}]0,\infty[\mid f \in S\}$. If (M,S) is a smooth pair then (a) for $f \in S$, $a \in R$, $f^{-1}]a,\infty[$, $f^{-1}]-\infty,a[$ are in $\mathcal{U}_S$, and (b) if $\mathcal{G} \subset \mathcal{U}_S$ then $\cap \mathcal{G} \in \mathcal{U}_S$ ($\cup \mathcal{G} \in \mathcal{U}_S$) if $\mathcal{G}$ is finite (locally finite).

3.1.4 Lemma

If (M,S) is a smooth pair then for any open covering $\{U_n\}_{n\geq 1}$ of M, where $U_n \in \mathcal{U}_S$, there is a locally finite cover $\{V_n\}_{n\geq 1} \subset \mathcal{U}_S$ such that $V_n \subset U_n$.

Proof

Let for each $n \geq 1$, $f_n \in S$ be such that $U_n = f_n^{-1}]0,\infty[$. Let $V_n = U_n \cap f_n^{-1}]-\infty,1/n[$. Now if $x \in U_i$, consider $G_x = f_i^{-1}]1/2\, f_i(x),\infty[$. Clearly G_x is an open set, $x \in G_x$, and G_x intersects V_n iff $n \leq \mathrm{Max}(\frac{2}{f_i(x)}, \frac{1}{n})$. Further V_n is a covering of M. This completes the proof of the lemma.

By the support of a real valued function f on a set X is meant the set $\{x \mid f(x) \neq 0\}$ in this chapter.

3.1.5 Definition

A Banach space E is S-smooth if there is a nontrivial function f of bounded support belonging to $S(E,R)$. A Banach space E is said to admit a S-partition of unity if for each open covering $\mathcal{G}$ of E, there exists a partition of unity $\{f_\alpha\}$ with (i) $f_\alpha \in S(E,R)$, (ii) support of f_α, for each α, in a subset of some set $V_\beta \in \mathcal{G}$, and (iii) for each point $x \in E$, there is a neighborhood U_x of x, such that all but finitely many of the functions f_α vanish on U_x.

3.1.6 Proposition

If E is S-smooth, then the norm topology of E is determined by $S(E,R)$, and conversely.

Proof

Let E be S-smooth. Since translates and homotheties are C^∞-maps, it follows that for each open set U of E and $x \in U$, there exists a function $g \in S(E,R)$ such that $g(x) \neq 0$ and the support of $g \subset U$. Now to complete the proof of the converse part let the norm topology of E and the topology determined by $S(E,R)$ be equivalent. Thus if U is the open unit ball of E, there are functions $f_i \in S(E,R)$, pairs of real numbers (a_i,b_i), $1 \leq i \leq n$, such that $\bigcap_{i=1}^{n} f_i^{-1}]a_i,b_i[\subset U$. Let g be a function in $C^\infty(R^n,R)$ such that $g(t_1,t_2,\ldots t_n) > 0$ if $a_i < t_i < b_i$, $1 \leq i \leq n$, and $g = 0$ otherwise. Then the map $\psi = g \circ (f_1 \times f_2 \times \ldots \times f_n) \circ d \in S(E,R)$, and clearly the support of $\psi \subset U$, completing the proof of the proposition.

3.1.7 Remark

It follows from the definition of S-categories that if a Banach space E is isomorphic with a Banach space $(F, \| \ \|)$, where $\| \ \| \in S(F\sim\{0\}, R)$, then E is S-smooth.

3.1.8 Theorem

If E is a S-smooth separable Banach space, then E admits S-partition of unity.

Proof

Let $\mathcal{G}$ be an open covering of E. For each $x \in E$, choose a $G_x \in \mathcal{G}$ such that $x \in G_x$. By the previous proposition, there exists an $f_x \in S(E,R)$, $f_x \geq 0$, such that (1) $f_x(x) > 0$, (2) support of $f_x \subset G_x$, and (3) if $\phi \in C^\infty(R,R)$, then $\phi \circ f_x \in S(E,R)$. Thus it follows that there are open sets V_x^1, V_x^2, $V_x^1 \subset \bar{V}_x^1 \subset V_x^2 \subset G_x$, and a function $h_x \in S(E,R)$, $0 \leq h_x \leq 1$, $h_x \equiv 1$ on $\bar{V}_x^1$, and $h_x \equiv 0$ outside V_x^2. Since E is second countable there is a sequence $x_i \in E$, such that $\{V_{x_i}^1\}$ is a covering of E. Denoting $V_{x_i}^1$, and h_{x_i} respectively by V_i^1 and h_i, $i \geq 1$, define functions g_i such that

$g_1 = h_1$, and

$$g_n(x) = h_n(x) \prod_{i=1}^{n} (1 - h_i(x)),$$

for all $n \geq 2$, and $x \in E$. Since $S(E,R)$ is an algebra, $g_n \in S(E,R)$, for all $n \geq 1$. Further $\{g_n\}$ is a locally finite sequence of functions, for if $x \in E$ let $N(x)$ be the smallest integer such that $x \in V^1_{N(x)}$. Then from the choice of the functions h_x it is verified that $g_m(z) = 0$ for all $z \in V^1_{N(x)}$, and $m > N(x)$. Now from the identity

$$\sum_{i=1}^{n} g_i(x) = 1 - \prod_{i=1}^{n} (1 - h_i(x)), \quad \text{for all } n \geq 1,$$

it follows with $N(x)$ chosen as above, that for each $x \in E$,

$$\sum_{i\geq 1} g_i(x) = \sum_{i=1}^{N(x)} g_i(x) = 1 - \prod_{i=1}^{N(x)} (1 - h_i(x)).$$

Further since the support of each $g_i \subset$ support of h_{x_i}, it follows that the support of each $g_i \subset G_{x_i} \in \mathcal{G}$. Thus the partition of unity $\{g_i\}_{i\geq 1}$, satisfies all the requirements.

3.1.9 Definition

If A is an arbitrary nonempty set, $c_0(A)$ is the Banach space of real valued functions f on A such that for each $\epsilon > 0$, the set $\{x \mid |f(x)| \geq \epsilon\}$ is finite, equipped with the norm $\|f\| = \sup_{x \in A} |f(x)|$. $c_0(A)$ is separable if and only if card $A \leq \aleph_0$.

We obtain the following corollary from the preceding theorem, noting that $c_0(N)$, ℓ_{2n}, $L_{2n}(\lambda)$ (λ is the Lebesgue measure on $[0,1]$) are C^∞-smooth.

3.1.10 Corollary

If E is any one of the separable Banach spaces c_0, ℓ_{2n}, $L_{2n}(\lambda)$, n is a positive integer, then E admits C^∞-partition of unity.

We note, that in the proof of the preceding theorem, separability of the Banach space E has been used in an essential manner. For this reason this proof does not extend whenever the space E is a nonseparable space, where E is any one of the spaces $c_0(A)$, $\ell_{2n}(A)$, or $L_{2n}(\lambda)$ where λ is an arbitrary positive measure on some measurable space. For this reason Torunczyk [62] developed a method to exhibit C^∞-partitions of unity subordinated to open covers of E, where E is any one of the preceding spaces.

3.1.11 Definition

Let S_0 be the set of all C^∞-functions $f: c_0(A) \to R$ which locally depend only on finitely many coordinates. More precisely S_0 is the set of functions $f: c_0(A) \to R$ such that given any $y \in c_0(A)$ there is a finite set $\{\alpha_1, \alpha_2, \ldots, \alpha_n\} \subset A$ and $\phi \in C^\infty(R^n)$ such that

$$f((x_\alpha)) = \phi(x_{\alpha_1}, x_{\alpha_2}, \ldots, x_{\alpha_n})$$

for all $x = (x_\alpha) \in c_0(A)$, in a neighborhood of y. Note that $(c_0(A), S_0)$ is a smooth pair, see definition 3.1.2(a).

3.1.12 Lemma

If $r > 0$ then the ball $B_r = \{y \in c_0(A) \mid \|y\| < r\}$ is in $\mathcal{U}_{S_0}$.

Proof

Let $\phi_r \in C^\infty(R,[0,1])$ such that $\phi_r(t) = 1$ for $t < \frac{r}{2}$, and $\phi_r(t) = 0$ iff $t \geq r$. Let f_r be defined on $c_0(A) \to R$ by setting $f_r(x) = \prod_{\alpha \in A} \phi_r(x_\alpha)$ if $x = \{x_\alpha\}_{\alpha \in A}$. It is verified that $f_r \in S_0$, and $B_r = f_r^{-1}]0,\infty[$, completing the proof of the lemma.

Before proving that $c_0(A)$ admits S_0-partitions of unity, we note the following general result (involving the topology of a Banach space E) on the existence of S-partitions of unity.

3.1.13 Lemma

Let (M,S) be a smooth pair. Then the following three statements are equivalent.

(1) M admits S-partitions of unity.

(2) For any closed set $A \subset M$ and an open neighborhood W of A there is some $U \in \mathcal{U}_S$, (For the definition of $\mathcal{U}_S$, see Remark 3.1.3.) with $A \subset U \subset W$.

(3) There exists a σ-locally finite base $\mathcal{G}$ for the topology of M with $\mathcal{G} \subset \mathcal{U}_S$.

Proof

Since M is paracompact, the only nontrivial part is the implication (3) ⇒ (2). Let A and W satisfy, and $\mathcal{G} \subset \mathcal{U}_S$ be a σ-locally finite base for the topology of M. Let $\mathcal{G} = \bigcup_{n \geq 1} \mathcal{G}_n$, with $\mathcal{G}_n$ locally finite for all $n \geq 1$. Let $\mathcal{U}_n = \{G \mid G \in \mathcal{G}_n,\ G \subset W\}$. Let $\mathcal{W}_n = \{G \mid G \in \mathcal{G}_n,\ G \cap A = \emptyset\}$, $X_n = \bigcup \mathcal{U}_n$, and $Y_n = \bigcup \mathcal{W}_n$. It follows from Remark 3.1.7 that $X_n \in \mathcal{U}_S$, and $Y_n \in \mathcal{U}_S$ for all $n \geq 1$. Since $\mathcal{G}$ is a base for the topology of M, $\{X_n\} \cup \{Y_n\}$ is a cover of M. Thus, as a consequence of Lemma 3.1.4, there exists sets $X_n^1 \subset X_n$, $Y_n^1 \subset Y_n$, such that $X_n^1 \in \mathcal{U}_S$, $Y_n^1 \in \mathcal{U}_S$ for all $n \geq 1$ and $\{X_n^1\}_{n\geq 1} \cup \{Y_n^1\}_{n \geq 1}$ is a locally finite cover of M. Using once again Remark 3.1.7 it follows that $A \subset X \subset W$, completing the proof of the lemma.

3.1.14 Theorem

$c_0(A)$ admits S_0-partitions of unity.

Proof

As seen from the preceding lemma it is enough to show that there is a σ-locally finite base $\mathcal{G}$ for the topology of $c_0(A)$, such that $\mathcal{G} \subset \mathcal{U}_{S_0}$.

We divide the proof of the existence of $\mathcal{G}$ with the above properties into two parts. In part 1 we define a set $\mathcal{G} \subset \mathcal{U}_{S_0}$. In part 2 we verify the set $\mathcal{G}$ has indeed the desired properties.

1. Let Q be the set of rationals. Let H_n be the set of all injections on $\{1,2,\ldots,n\}$ into A, and

$K_n = \{(x,y) | (x,y) \in Q^n \times Q: \inf_{1 \le i \le n} |x_i| > y > 0\}$ where $x = (x_1, x_2, \ldots, x_n)$. For each $\sigma \in H_n$, let T_σ be the linear operator, $T_\sigma : R^n \to c_0(A)$ defined by $T_\sigma(x) = \{f_\alpha\}_{\alpha \in A}$, where $f_\alpha = 0$ if $\alpha \notin$ range σ, and $f_\alpha = x_i$ if $\alpha = \sigma(i)$. Define now

$$\mathcal{G} = \{B_{1/n}\}_{n \ge 1} \cup \bigcup_{n \ge 1} \{T_\sigma(x) + B_y, (x,y) \in K_n, \quad \sigma \in H_n\}.$$

2. First we observe that if $f = \{f_\alpha\} \in c_0(A)$, and $\epsilon > 0$, then there is a set $W \in \mathcal{G}$ such that $f \in W$, and diam $W < 2\epsilon$. To verify this assertion we may assume $\|f\| > \epsilon$, since the balls $B_{1/n} \in \mathcal{G}$, for $n \ge 1$. Since the set $\{|f_\alpha|\} \cup \{0\}$ is compact and countable there is a rational number y, $0 < y < \epsilon$, such that $y \ne |f_\alpha|$ for any $\alpha \in A$. Let $\{\alpha_1, \alpha_2, \ldots, \alpha_n\}$ be the subset of A, such that $|f_{\alpha_i}| > y$, $1 \le i \le n$, and let $x = (x_1, x_2, \ldots, x_n) \in Q^n$ satisfy $|x_i| > y$, and $|x_i - f_{\alpha_i}| < y$, and let σ be the injection on $\{1,2,\ldots,n\} \to A$ defined by the map $h(i) = \alpha_i$. Then $(x,y) \in K_n$, and $\|T_\alpha(x) - f\| < y$. Thus $x \in W = T_\sigma(x) + B_y$, and diam $W < 2\epsilon$.

Now we proceed to verify that $\mathcal{G}$ is locally finite. Since $\{B_{1/n}\}_{n \ge 1}$ and $\bigcup_{n \ge 1} K_n$ are countable sets it is enough to show for all $n \ge 1$, and $(x,y) \in K_n$, the family $\{T_\sigma(x) + B_y : \sigma \in H_n\}$ is locally finite. Fix $n \ge 1$, and $(x,y) \in K_n$ and $f \in c_0(A)$. Let $\delta = \inf_{1 \le i \le n} \{|x_i| - y\}$, and $P = \{\alpha_1, \alpha_2, \ldots, \alpha_n\}$, be as in the preceding paragraph. Now for any $a = \{a_\alpha\} \in f + B_\epsilon$ $|a_\alpha| < 2\epsilon$ if $\alpha \notin P$. If $z = \{z_\alpha\} \in T_\sigma(x) + B_y$, the inequality $|z_\alpha| > 2\epsilon$ is true for $\alpha \in$ range σ. Thus the ball $f + B_\epsilon$ intersects only sets $T_\sigma(x) + B_y$ for which range $\sigma \subset P$. Clearly such sets are finite in number.

3.1.15 <u>Theorem</u>

Let (M,S) be a smooth pair. Then M admits S-partitions of unity iff there is a homeomorphism $u: M \to c_0(A)$, for some set A such that $p_\alpha \circ u \in S$, for $\alpha \in A$.

Proof

Let (M,S) be such that there is a homeomorphism $U:M \to c_o(A)$ with $p_\alpha \circ u \in S$. Let $\mathcal{G}$ be the base of $c_o(A)$ constructed in the preceding theorem. Let $\mathcal{G}_1 = u^{-1}(\mathcal{G})$. Then $\mathcal{G}_1$ is a σ-locally finite base for the topology of X. Further the conditions II and III in the definition 3.1.2(a) of a smooth pair, in the presence of the property $p_\alpha \circ u \in S$, assures that $f \circ u \in S$ if $f \in S$. Thus $\mathcal{G}_1 \subset \mathcal{U}_S$, completing the proof of "if part" in the theorem.

Next let (M,S) admit S-partitions of unity. Then Lemma 3.1.13 assures the existence of σ-locally finite base $\mathcal{V} \subset \mathcal{U}_S$ for the topology of M. Let $\mathcal{V} = \bigcup_{n \geq 1} \mathcal{V}_n$, where each $\mathcal{V}_n$ is locally finite, and $\mathcal{V}_n \cap \mathcal{V}_m = \emptyset$ if $m \neq n$. Choose for each $V \in \mathcal{V}$ a function $f_V \in S$ such that $f_V : M \to [0,1]$ and $f_V^{-1}]0,1] = V$. Define $u:M \to c_o(\mathcal{V})$ by setting $p_V \circ u(x) = \frac{1}{n} f_V(x)$ if $x \in \mathcal{V}_n$. Let W be an open set such that $W \cap V \neq \emptyset$, for $V \in \{V_i\}_{i=1}^k$ where $\{V_i\}_{i=1}^k \subset \bigcup_{i=1}^n \mathcal{V}_i$. u is continuous since $||u(y) - u(x)|| \leq \frac{1}{n}$ for all $y \in W \cap f_V^{-1}(\xi_x - 1/2n, \xi_x + 1/2n)$ where $\xi_x = f_V(x)$. Since if $x \in V \in \mathcal{V}_n$, and $y \notin V$ imply $||u(x) - u(y)|| \geq \frac{1}{n} f_V(x)$, $u:M \to u(M) \subset c_o(A)$ is a homeomorphism. This completes the proof of the "only if part".

We deduce from this theorem several results assuring the existence of S-partitions of unity, S suitably interpreted, for certain Banach spaces.

3.1.16 Corollary

Let E, F, be normed linear spaces with F admitting C^n-partitions of unity. If there exists a homeomorphism $U:E \to F$ with $u \in C^n(E,F)$, then E also admits C^n-partitions of unity.

3.1.17 Corollary

If E_1, E_2 are two normed linear spaces admitting C^k-partitions of unity then $E_1 \times E_2$ admits C^k-partitions of unity.

Let $u_1:E_1 \to c_o(A_1)$, $u_2:E_2 \to c_o(A_2)$ be homeomorphisms assured by Theorem 3.1.15. Then if $u:E_1 \times E_2 \to c_o(A_1) \times c_o(A_2)$ is the map defined by $u(e_1,e_2) = L(u_1(e_1), u_2(e_2))$ where $L: c_o(A_1) \times c_o(A_2) \to c_o(A_1 \times A_2)$ is the natural isomorphism, u satisfies the conditions stated in Theorem 3.1.15, completing the proof.

3.1.18 Corollary

Theorem 3.1.14 yields at once that $c_o(A)$ admits C^∞-partitions of unity.

3.1.19 Theorem

Every Hilbert space H admits C^∞-partitions of unity.

Proof

Let $H = \ell_2(A)$ where $1 \notin A$. Let $u:\ell_2(A) \to c_o(\{1\} \cup A)$ be defined by

$$p_\beta \circ u((f_\alpha)) = \begin{cases} \|f_\alpha\|^2 & \text{for } \beta = 1 \\ f_\beta & \text{for } \beta \in A . \end{cases}$$

It is verified that u is a homeomorphism. The result follows from Theorem 3.1.14.

The next lemma is useful to discuss the existence of S-partitions of unity on reflexive Banach spaces. Since the proof follows from well known properties of reflexive spaces, and techniques similar to those adopted in the preceding theorems we do not present the proof. However, we refer for a complete proof to section III page 9 in [62] .

3.1.20 Lemma

Let E be a reflexive Banach space whose norm is locally uniformly convex and let $L:E \to c_o(A)$ be a continuous 1-1 linear map. Suppose $A \cap N = \emptyset$. If $\phi_n \in C^\infty(R)$ $(n \geq 1)$ are nondecreasing functions with $\phi_n(t) = t$ for $t > \frac{1}{n}$, and $\phi_n(t) = 0$ for $t \leq \frac{1}{n}$, then the map $u:E \to c_o(N \cup A)$ defined by

$$p_\beta \circ u(x) = \begin{cases} \phi_\beta\left(\frac{|||x|||}{\beta}\right), & \beta \in N \\ p_\beta \circ L(x) , & \beta \in A , \end{cases}$$

is a homeomorphic embedding.

3.1.21 Corollary

If E is a reflexive Banach space admitting an equivalent norm which is locally uniformly convex, and of class C^n, away from 0, then E admits C^n-partitions of unity.

The corollary follows at once combining the preceding lemma with theorem 3.1.15.

We stated in Chapter II the smoothness properties of the norm in $L_p(\mu)$ spaces. Theorem 3.1.15 together with the preceding lemma, applied to these spaces yield the following results.

3.1.22 Corollary

For every integer $n \geq 1$, the Banach spaces $L_{2n}(\mu)$ admit C^∞-partitions of unity. If $1 < p$, then $L_p(\mu)$ admits C^m-partitions of unity, $1 \leq m < p$.

3.1.23 Remark

It is known that every reflexive space admits an equivalent locally uniformly convex norm of class C^1 away from 0, see [11]. Thus by arguments similar to the above it follows that every reflexive Banach space admits C^1-partitions of unity.

3.2 A nonlinear characterization of superreflexive spaces

Wells [64] observed that the Banach space c_o does not admit a nontrivial uniformly continuously differentiable function with bounded support. Aron [3] proved the same result for the Banach spaces $C(X)$, X an infinite compact Hausdorff space. We establish in this section, among others, a complete characterization of Banach spaces which admit nontrivial uniformly continuously differentiable functions with bounded support. The results as well as their proofs are taken from Sundaresan [59]. The theorems 3.2.11 and 3.2.16 have been obtained independently by John, Torunczyk and Zizler [68] by different techniques.

3.2.1 Lemma

Let E be a Banach space, and $f: E \to R$ be a uniformly continuously differentiable function. Then

(a) if U is a bounded subset of E, then $f|U$ is Lipschitzian i.e. there is a positive number M such that for $x,y \in U$, $|f(x) - f(y)| \leq M \|x-y\|$.

(b) if the support of f is bounded, then f is Lipschitzian on all of E.

Proof

Let $B_r(0)$ be an open ball, center 0 and radius r such that $B_r(0) \supset U$. If f is uniformly continuously differentiable then

$$\sup_{x \in B_r(0)} \|Df(x)\| < \infty .$$

If M is the preceding supremum, then by the mean value theorem it follows that $|f(x) - f(y)| \leq M \|x-y\|$.

3.2.2 Definition

A Banach space E is said to be U^1-smooth if there exists a nonzero uniformly continuously differentiable real valued function f such that the support of $f = \{x | f(x) \neq 0\}$ is bounded.

3.2.3 Lemma

If E is a U^1-smooth Banach space and λ is a positive real number, then there is a uniformly continuously differentiable real-valued function f on E with $f(0) = 1$ and $f(x) = 0$ if $\|x\| \geq \lambda$.

The lemma is a consequence of the fact that translates, and homotheties are uniformly continuously differentiable functions on E into E.

The composites, and products of uniformly continuously differentiable functions are in general not uniformly continuously differentiable. This remark motivates the next two lemmas. As the results are well-known, proofs are not supplied here.

3.2.4 Lemma

If f, g are two uniformly continuously differentiable real valued functions on a Banach space E, and the support of f, or g is bounded then fg is uniformly continuously differentiable.

3.2.5 Lemma

If E, F, G are three Banach spaces $f:E \to F$ and $g:F \to G$ are two uniformly continuously differentiable functions such that the derivatives Df, Dg are bounded mappings on $E \to L(E,F)$, and on $F \to L(F,G)$ respectively then their composite $g \circ f:E \to G$ is uniformly continuously differentiable.

3.2.6 Lemma

If f is a nontrivial uniformly continuously differentiable real valued function on a Banach space E with its support, U_f, bounded, then there is a uniformly continuously differentiable real valued fucntion g on E with $0 \leq g \leq 1$, with its support $U_g = U_f$.

3.2.7 Lemma

If E, f, U_f are as in lemma 3.2.6, and if $x \in U_f$, then there is a real valued uniformly continuously differentiable function g on E such that the support of $g = U_f$, $0 \leq g \leq 1$, and $g \equiv 1$ in a closed neighborhood of x.

3.2.8 Theorem

If E is U^1-smooth, then every ultrapower $E(S,\Gamma)$ of E is U^1-smooth.

Proof

Let the norms of E, and $E(S,\Gamma)$ be respectively $\| \ \|$, and $||| \ |||$. Since E is U^1-smooth there is a uniformly continuously differentiable function f, $f \neq 0$, on $E \to R$ with its support in the unit ball of E, see Lemma 3.2.3. Let $\tilde{x} \in E(S,\Gamma)$, and $\{x(s)\}_{s\in S}$ be a representative of $\tilde{x}$. Since f is bounded $\lim_{\Gamma} f(x(s))$ exists. Further uniform continuity of f assures that the preceding limit is independent of the choice of $\{x(s)\}$ from $\tilde{x}$. Let $f^*:E(S,\Gamma) \to R$ be defined by setting $f^*(\tilde{x}) = \lim_{\Gamma} f(x(s))$. Since the support of f is in the unit ball of E, it is verified from the definition of $|||\ \ |||$ that the support of f^* is in the unit ball of $E(S,\Gamma)$.

Since $Df:E \to E^*$ is a uniformly continuous mapping with bounded range, it is verified that if $\tilde{x}$, $\tilde{y} \in E(S,\Gamma)$, and if $\{x(s)\}$, $\{y(s)\}$ are a pair of representatives of $\tilde{x}$, $\tilde{y}$, then $\lim_{\Gamma} Df(x(s))(y(s))$ is independent of the representatives. For $\tilde{x}$, $\tilde{y} \in E(S,\Gamma)$ define $\ell_{\tilde{x}}(\tilde{y}) = \lim_{\Gamma} Df(x(s))(y(s))$. It is verified that $\ell_{\tilde{x}} \in (E(S,\Gamma))^*$, since Df is bounded. Now if $\tilde{h} \in E(S,\Gamma)$, and $\{h(s)\} \in \tilde{h}$, then $f^*(\tilde{x} + \tilde{h}) = \lim_{\Gamma} f(x(s) + h(s)) = \lim_{\Gamma} \{f(x(s)) + Df(x(s))(h(s)) + {}$ $+ \theta_{x(s)}(h(s))\}$, where it is noted that since f is uniformly continuously differentiable, that if $\epsilon > 0$, there is a $\delta > 0$, such that $|\theta_x(y)| \leq \epsilon \|y\|$, if $\|y\| \leq \delta$, for all $x \in E$. Let now $|||\tilde{h}||| \leq \delta$. Then there is a set $J \in \Gamma$, such that for all $s \in J$, $|\theta_{x(s)}(h(s))| \leq \epsilon \|h(s)\|$. Hence $\lim_{\Gamma} |\theta_{x(s)}(h(s))| \leq \epsilon |||\tilde{h}|||$ if $|||\tilde{h}||| \leq \delta$, and f^* is differentiable at $\tilde{x}$ with $Df^*(\tilde{x}) = \ell_{\tilde{x}}$.

Since $Df:E \to E^*$ is uniformly continuous map on $E \to E^*$, it is verified that the map $Df:E(S,\Gamma) \to (E(S,\Gamma))^*$ is uniformly continuous, once again working with suitable members of Γ as has been done above. Thus $E(S,\Gamma)$ is U^1-smooth.

3.2.9 <u>Remark</u>

Before proceeding to the main result here, let us note that the preceding theorem implies at once, that if E is U^1-smooth, $F \ll E$, then F is U^1-smooth. Further if E is superreflexive, then since E is isomorphic with a uniformly smooth Banach space, it follows that E is U^1-smooth.

3.2.10 Theorem

If E is U^1-smooth then E is superreflexive.

Proof

Let $0 < \theta < 1$. Lemma 3.2.3 assures that there is a U.C.D. real-valued function f on E such that $f(0) = 1$, and $f(x) = 0$ if $\|x\| \geq \frac{\theta}{4}$. Since f is U.C.D. if $0 < \epsilon < 1$, there is a positive integer M such that if $h \in E$, $\|h\| \leq \frac{1}{M}$, then

(a) $\quad |f(x + h) - f(x) - Df(x)(h)| \leq \epsilon \|h\|$.

If possible let E be nonreflexive. Then by a theorem of James, see Theorem 7 in [27], it follows that there is a set X containing the set W of positive integers, and a subspace L of the Banach space $B(X)$ of bounded real-valued continuous functions on X with the supremum norm, isometric with E, admitting a sequence $\{z_n\}_{n\geq 1}$, such that for $n \geq 1$

$$z_n(i) = \theta, \quad 1 \leq i \leq n, \quad i \in W,$$

$$z_n(i) = 0, \quad 1 > n, \quad i \in W,$$

and

$$|z_n(t)| \leq 1 \quad \text{for} \quad t \in X \sim W.$$

Let $x_{n,0} = \frac{1}{2} z_n$, $x_{0,n} = -\frac{1}{4} z_n$ for $n \geq 1$, and $x_{n,k} = \frac{3}{4} z_n - \frac{1}{4} z_{n+k}$ if $n \geq 1$, $k \geq 1$. Clearly $\|x_{n,k}\| \leq 1$ for all the pairs of integers (n,k) for which $x_{n,k}$ is defined. Consider the polygonal path $P \subset L$ defined by

$$P = \bigcup_{i=0}^{2^M-1} [x_{2^M-i,i}, x_{2^M-i-1,i+1}]$$

where M is the positive integer chosen to satisfy the inequality (A) in the preceding paragraph. Consider the derivative $Df(0)$ of f at 0. By our choice of $x_{n,k}$, $Df(0)(x_{2^M,0}) = 0$ if and only if $Df(0)(x_{0,2^M}) = 0$, and $Df(0)(X_{2^M,0})$ is positive (negative) if and only if $Df(0)(x_{0,2^M})$ is negative (positive). Since P is connected there is a $\xi \in P$ such that $Df(0)(\xi) = 0$. If

$$\xi \in [x_{2^M - i_0, i_0}, x_{2^M - i_0, i_0+1}],$$

then

$$\xi(j) = \frac{1}{2}\theta \quad \text{if} \quad 1 \le j \le 2^M - i_0 - 1, \quad j \in W,$$
$$\xi(j) = -\frac{1}{4}\theta, \quad \text{if} \quad 2^M - i_0 + 1 \le j \le 2^M, \quad j \in W,$$
$$\xi(j) \in [-\frac{1}{4}\theta, \frac{1}{2}\theta] \text{ if } j = 2^M - i_0, \quad j \in W,$$

and

$$\|\xi\| \le 1.$$

Thus if $\xi(j_0) < \frac{1}{2}\theta$ for some $j_0 \in W$, $1 \le j_0 \le 2^M$ (which is the case if $\xi(j_0) \in]-\frac{\theta}{4}, \frac{\theta}{2}[$ or $\xi(j_0) = -\frac{\theta}{4}$), then $\xi(j) = -\frac{1}{4}\theta$ for all $j \in W$, $j_0 + 1 \le j \le 2^M$. Now if $2^M - i_0 - 1 \ge 2^{M-1}$, choose $\xi_1 = \frac{\xi}{M}$, otherwise $\xi_1 = -\frac{\xi}{M}$. The ξ_1 thus chosen has the properties $\|\xi_1\| \le \frac{1}{M}$, $Df(0)(\xi_1) = 0$, and $\xi_1(j) \ge \frac{\theta}{4M}$ for at least 2^{M-1} values of $j \in W$, $1 \le j \le 2^M$.

Next consider the derivative $Df(\xi_1)$. Then as before there is a $\xi' \in P$ such that $Df(\xi_1)(\xi') = 0$. From the properties of ξ noted in the preceding paragraph, since $\xi_1 = \pm\frac{\xi}{M}$, the restriction of ξ_1 to the set $Q = \{j \mid 1 \le j \le 2^M\} \subset W$, has range either in the set $\{\frac{\theta}{2M}, -\frac{\theta}{4M}\}$ or $\{-\frac{\theta}{2M}, \frac{\theta}{4M}\}$ except possibly for one value of $j \in Q$. These observations imply either (i) $(\xi_1 + \frac{\xi'}{M})(j) \ge \frac{2}{4M}$ or (ii) $(\xi_1 - \frac{\xi'}{M})(j) \ge \frac{2\theta}{4M}$ for at least 2^{M-2} integers $j \in Q$. Let $\xi_2 = \frac{\xi'}{M}$ or $-\frac{\xi'}{M}$ according as (i) or (ii) is the case. Repeating this procedure inductively it follows that there is a sequence $\{\xi_i\}_{i=1}^M$ in L such that $\|\xi_i\| \le \frac{1}{M}$, $Df(\sum_{i=1}^{k-1} \xi_i)(\xi_k) = 0$, $\sum_{i=1}^{k} \xi_i(j) \ge \frac{k\theta}{4M}$ for $1 \le k \le M$, for at least 2^{M-k} values of $j \in W$. From the choice of $f, M, \{\xi_i\}_{i=1}^M, \epsilon$, together with the inequality $\|\sum_{i=1}^{k} \xi_i\| \ge \frac{k\theta}{4M}$ it follows that

$$1 = |f(\sum_{i=1}^{M} \xi_i) - f(0)| \le \sum_{k=1}^{M} |f(\sum_{i=1}^{k} \xi_i) - f(\sum_{i=1}^{k-1} \xi_i) - Df(\sum_{i=1}^{k-1} \xi_i)(\xi_k)|$$

$$\le \sum_{k=1}^{M} \epsilon \|\xi_i\| \le \epsilon < 1,$$

a contradiction, completing the proof of the theorem.

Combining the two theorems stated above we obtain the following non-linear characterization of a superreflexive Banach space.

3.2.11 Theorem [Sundaresan]

A Banach space E is superreflexive iff it is U^1-smooth.

3.2.12 Corollary

If E is either the Banach space c_0 or $C(X)$, X an infinite compact Hausdorff space then E is not U^1-smooth.

Before proceeding to a characterization of superreflexive spaces in terms of partitions of unity, we prove a lemma and state a result from Nemirovskii and Semenov [46] without proof.

3.2.13 Lemma

If E is a uniformly smooth Banach space, and $\epsilon > 0$ there exists a uniformly continuously differentiable function f with $f \equiv 1$ on the unit ball $B_1(0)$, and $f \equiv 0$ outside the ball $B_{1+\epsilon}(0)$.

Proof

Since E is U^1-smooth there exists a uniformly continuously differentiable function $f_0 : E \to R$ such that $0 \leq f_0 \leq 1$, $f_0 = 0$ outside $B_1(0)$, $f_0 = 1$ in a neighborhood V of the origin. Let $\delta > 0$ be such that $2\delta < \min(\epsilon, 1)$, and $B_{2\delta}(0) \subset V$. Let $\alpha : R \to R$ be a C^1-function, such that $0 \leq \alpha \leq 1$, $\alpha(r) = 1$ if $2\delta \leq r \leq 1$, $\alpha(r) = 0$ if $r \leq \delta$ or $r \geq 1 + \delta$. Consider the function f_1 on $E \to R$ defined by $f_1(x) = \alpha(\|x\|)$. Since the norm, $\| \ \|$, of E is uniformly smooth, the $\| \ \|$ is uniformly continuously differentiable in regions $R(\lambda, \mu)$. Hence f_1 is a uniformly continuously differentiable real valued function with support $f_1 \subset B_{1+\delta}(0)$. Now if $\beta : R \to R$ is a C^1-function, with $0 \leq \beta \leq 1$, $\beta = 1$ on $[1, 2]$, $\beta(r) = 0$ if $r \leq \frac{1}{2}$ or $r \geq \frac{5}{2}$, it is verified that the function $\beta(f_0 + f_1)$ satisfies all the requirements, completing the proof.

3.2.14 Lemma

If E is a uniformly convex and uniformly smooth Banach space, then the restrictions of the uniformly continuously differentiable functions on E to any closed ball $\overline{B_r(0)}$ are dense in the space of uniformly continuous functions on $\overline{B_r(0)}$.

For a proof of this lemma, see Remark 4.2.12.

3.2.15 Proposition

If G is an open subset of a superreflexive Banach space E, then there is a uniformly continuously differentiable (U.C.D.) function, f, on $E \to R$ such that $0 \leq f \leq 1$, $f \neq 0$, and support of $f = G$.

Proof

As a primary case let G be a bounded open subset of E. Let $C = E \sim G$, and $g: E \to R$ be the function $f(x) = d(x,C)$, the distance of x from C. Then g is a uniformly continuous real valued function. Let r be a real number such that $G \subset \overline{G} \subset B_{r/2}(0) \subset \overline{B_r(0)}$. Consider the restriction of g to $\overline{B_r(0)}$. By the preceding lemma there is a U.C.D. function f_n on $E \to R$ such that $\sup_{x \in \overline{B_r(0)}} |f_n(x) - g(x)| < \frac{2}{n}$, for $n \geq 1$. Let $\phi_n: E \to R$ be a U.C.D. function such that $0 \leq \phi_n \leq 1$, $\phi_n \equiv 1$ on $B_{r/2}(0)$, and $\phi_n \equiv 0$ outside $B_r(0)$. Consider $h_n = \phi_n f_n$. Now if $\alpha_n: R \to R$ is a C^1-function with $\alpha_n(t) = 0$ if $t \leq 1/n$ or $t \geq 2/n$, $0 \leq \alpha_n \leq 1$, let $g_n = \alpha_n(f_n)$. Thus $g_n(x) = 0$ if $0 \leq g(x) \leq 1/n$, $0 \leq g_n \leq 1$, and $g_n(x) = 0$ if $g(x) \geq 4/n$. Let $f(x) = \sum_{n \geq 1} \frac{1}{2^n} g_n$. It is verified that f is U.C.D. real valued function on E, support of $f = G$, $0 \leq f \leq 1$.

Now let G be an arbitrary open set, and $G_n = B_n(0) \cap G$, $n = 1,2,3,\ldots$. Corresponding to each n, there exists a U.C.D. real valued function f_n on E, such that $0 \leq f_n \leq 1$, support $f_n = G_n$. Consider $f = \sum_{n \geq 1} \frac{1}{2^n} f_n$. f is the desired function.

3.2.16 Theorem

If E is a superreflexive Banach space, and $\mathcal{G}$ is an open covering of E, there exists a locally finite family of U.C.D. functions $\{f_\alpha\}$, which is a partition of unity subordinated to $\mathcal{G}$.

Proof

Let $\{\mathcal{G}_n\}_{n\geq 1}$ be a sequence of discrete families of open sets such that $\bigcup_{n\geq 1} \mathcal{G}_n$ is a refinement of $\mathcal{G}$. Let $S_n = \bigcup\{G | G \in \mathcal{G}_n\}$. Let g_n be a U.C.D. function, $0 \leq g_n \leq 1$, on E such that support of $g_n = S_n$. For each $U \in \mathcal{G}_n$, let $f_{n,U}$ be a U.C.D. function, $0 \leq f_{n,U} \leq 1$, such that the support of $f_{n,U} = U$. Now for each pair (n,j), $n \geq 1$, $j \geq 1$, let $f_{n,j}$ be a U.C.D. function, $0 \leq f_{n,j} \leq 1$, such that $f^{-1}_{n,j}(0) = g_n^{-1}(0)$, $f^{-1}_{n,j}(1) = g_n^{-1}[\frac{1}{j},1]$. Let $\sigma : N \to N \times N$ be an onto bijection on the set N of positive integers, and $\sigma(n) = (i_n, j_n)$. Let $p_n = f_{i_n, j_n}$, $n \geq 1$. Let $h_n = \prod_{i=1}^{n} p_n(1-p_{n-i})$, where p_0 is the zero function. It is verified that the functions h_n, $n \geq 1$ are U.C.D. functions with $0 \leq h_n \leq 1$, using lemma 3.2.4. $\mathcal{S}_{i_n} = \{f_{n,U}\}\ U \in \mathcal{G}_{i_n}$. For $f \in \mathcal{S}_{i_n}$, define $h_{n,f}(x) = h_n(x)$, if $x \in$ support of f, and $h_{n,f}(x) = 0$ if $x \notin$ support of f. Since the set $\{\text{support } f | f \in \mathcal{S}_{i_n}\}$ is discrete it follows that $\{h_{n,f}\}_{f \in \mathcal{S}_{i_n}}$ are U.C.D. functions on $E \to [0,1]$.

We proceed to verify that $\{h_{n,f}\}_{\substack{f \in \mathcal{S}_{i_n} \\ n\geq 1}}$ is a locally finite partition of unity subordinated to the covering $\mathcal{G}$. Let $x \in E$. Choose smallest positive integer n such that $f_{i_n,j_n}(x) > 0$. Thus $h_n(x) > 0$. Pick $U \in \mathcal{G}_{i_n}$ such that $x \in U$. The function $h_{n,f_U}(x) > 0$, and the support $f_{n,f_U} \subset U \in \mathcal{G}_{i_n}$. Hence the $\{\text{support of } h_{n,f}\}_{\substack{n \geq 1 \\ f \in \mathcal{S}_{i_n}}}$ is a refinement of $\mathcal{G}$. If $x \in E$ there is a neighborhood V of x such that for some pair (i,j) of integers, $f_{i,j}(y) = 1$ for all $y \in V$. If $(i,j) = (i_k, j_k)$, then $p_k | V = 1$.

From this it follows that all but finitely many functions h_n vanish on V. Note from the definitions of h_n's, $1 - \prod_{i=1}^{k}(1 - p_i) = \sum_{i=1}^{k} h_i$ for all $k \geq 1$. Hence it follows that $\{h_n\}$ is a locally finite partition of unity. Now since $\{\text{support } f \mid f \in \mathcal{S}_{i_n}\}$ is discrete for all $n \geq 1$, it follows that $\Sigma h_{k,f} = h_k$, for each $k \geq 1$. Thus $\{h_{k,f}\}_{\substack{k \geq 1 \\ f \in \mathcal{S}_{i_k}}}$ is a locally finite partition of unity subordinated to $\mathcal{G}$, consisting of U.C.D. real valued functions on E.

3.3 Functions on Banach spaces with Lipschitz derivatives

We study in this section Banach spaces which admit nontrivial real valued functions f of class C^k with support bounded, and the kth derivative satisfying a Lipschitz condition.

3.3.1 Definition

If E, F are Banach spaces, then we define $B_M^k(E,F) = \{f \mid f \in C^k(E,F)$, and $\|D^k f(y) - D^k f(x)\| \leq M\|x-y\|$, for all $x,y \in E\}$, and $B^k(E,F) = \bigcup_{M>0} B_M^k(E,F)$.

A Banach space E is said to be B^p-smooth if there is a function $f \in B^p(E,R)$, such that the support of f is bounded. The concept of B^p-smoothness defines a S-category, see Section 3.1. From the results in Section 2.2, see 2.2.1 and 2.2.2 on the order of differentiability of the norm in the spaces $\ell_p(L_p)$, $1 < p < \infty$ it follows that if p is not an integer then these spaces are B^k-smooth where k is the largest integer less than p, and if p is an odd (even) integer then these spaces are B^{p-1}-smooth (B^k-smooth for all $k \geq 1$).

The theorem is much deeper than the information provided in the preceding paragraph, and we focus our attention in developing a proof of this theorem.

We state a lemma on functions of class B^k before stating the theorem.

3.3.2 Lemma

If E, F are two Banach spaces and $1 \leq k < \infty$, and $f \in B^k(E,F)$, and if m is the greatest integer less than or equal to k, then if $x \in E$ and $\epsilon > 0$, there is a $\delta > 0$ such that

$$\| f(x+h)-f(x)-\ldots-\frac{1}{m!} D^m f(x)(h^{(m)}) \| \leq \epsilon \|h\|^k \quad \text{if } \|h\| \leq \delta .$$

The lemma follows from Taylor's theorem.

3.3.3 Theorem [Bonic and Frampton]

If E is an infinite dimensional subspace of ℓ_p, and p not an even integer, then E is not D^p-smooth if p is an odd integer, and it is not C^n_{p-n}-smooth, if p is not an integer, and n is the largest integer $\leq p$. In particular it is not B^p-smooth.

Before proceeding to the proof some observations are necessary on polynomials on a ℓ_p-space. Note that if $x,y \in \ell_p$, with disjoint supports, then

(i) $\|x+y\|^p = \|x\|^p + \|y\|^p$, and

(ii) if $h_i \to 0$ weakly and $\|h_i\| = a$, for all $i \geq 1$, then $\|x+h_i\|^p \to \|x\|^p + a^p$, for every $x \in E$.

3.3.4 Lemma

If $n < p$, every polynomial of degree n on E is weakly sequentially continuous.

Proof

Since polynomials of degree 1 are affine real valued functionals on E, for $n = 1$ the lemma is verified. Thus the proof is reduced to verifying all monomials of degree $n > 1$, are weakly sequentially continuous if every polynomial of degree $< n$ is weakly sequentially continuous. Let $Q(x)$ be a monomial of degree n. Thus $Q(x+h) = Q(x) + A(x,h) + Q(h)$ where for fixed x, $A(x,h)$ is a polynomial of degree less than n in h, and $A(x,0) = 0$. By the induction assumption $A(x,h)$ is weakly sequentially continuous.

We proceed to verify that $Q(h_j) \to 0$ if $h_j \to 0$ weakly. If this is false there is a monomial P of degree n, ($P = Q$ or $-Q$), a sequence $h_j \to 0$ weakly, with $\|h_j\| = 1$, such that $P(h_j) \geq a$ for some positive number a, for all $j \geq 1$. As above let $P(x+h) = P(x) + B(x,h) + P(h)$, where for fixed x, $B(x,h)$ is a polynomial in h of degree less than n, with $B(x,0) = 0$. Clearly $B(x,h_j) \to 0$ for every x, by assumption. Let $n_1 = 1$, and $x_1 = h_{n_1}$. Choose n_j inductively so that $|B(x_{j-1},h_{nj})| < \frac{a}{2}$, and $\|x_{j-1}+h_{nj}\|^p < \|x_{j-1}\|^p + 2$, this being possible from the observation preceding the lemma. Let $x_j = x_{j-1} + h_{nj}$. Then $\|x_j\|^p \leq 2j$ and $P(x_j) = P(x_{j-1}) + B(x_{j-1},h_j) + P(h_j) \geq P(x_{j-1}) + \frac{1}{2}a$. Since $P(x_1) > a$, we have that $P(x_{j-1}) > j\frac{a}{2}$. Thus

$$\frac{P(x_j)}{\|x_j\|^n} \geq \frac{aj}{2(2j)^{n/p}} \to \infty \quad \text{as } j \to \infty, \quad \text{since } 1 - \frac{n}{p} > 0.$$

This contradicts the continuity of the n-linear form P.

3.3.5 Lemma

If p is an odd integer and E is an infinite dimensional subspace of ℓ_p, P a polynomial of degree p on E and $a > 0$, there is a sequence $\{h_j\}$ in E such that $\|h_j\| = a$, $P(h_j) \to P(0)$, and $\|z+h_j\|^p \to \|z\|^p + a^p$ for all $z \in E$.

Proof

Let $P(x) = Q(x) + R(x)$, where $R(x)$ is a monomial of degree p, and Q is a polynomial of degree less than p. Consider the connected set $C_n = \{x \in E, x_1 = \ldots = x_n = 0, \|x\| = a\}$. Since E is infinite dimensional, C_n is nonempty and since $R(x) = -R(-x)$, there is a $h_n \in C_n$, such that $R(h_n) = 0$. Now $\|h_n\| = a$, $h_n \to 0$ weakly and $\|z+h_n\|^p \to \|z\|^p + a^p$ for all $z \in E$, and the preceding lemma implies $P(h_n) = Q(h_n) \to Q(0) = P(0)$.

The lemmas 3.3.2, 3.3.4 and 3.3.5 together yield the following result which is of considerable importance in the proof of the theorem.

3.3.6 Lemma

Let E be an infinite dimensional subspace of ℓ_p, with p not an even integer. Let $x \in E$, and $\epsilon > 0$. Then there is a $\delta > 0$, and a sequence $h_n \in E$, $\|h_n\| = \delta$, $\|f(x+h_n)-f(x)\| \leq \epsilon\, \delta^p$ for every n and $\|z+h_n\|^p \to \|z\|^p + \delta^p$ for all $z \in E$.

Proof

It follows from Lemma 3.3.2 that there is a polynomial P of degree less than or equal to p and a $\delta > 0$ such that $P(0) = 0$ and $|f(x+h) - f(x) - P(h)| \leq \epsilon \frac{\delta^p}{2}$ if $\|h\| \leq \delta$. If p is an odd integer the result follows from Lemma 3.3.5. If p is not an odd integer pick a sequence $\{h_n\}$ in E so that $\|h_n\| = \delta$ and $\|h_n\| \to 0$ weakly. Then by Lemma 3.3.4 $P(h_n) \to 0$ and $|f(x+h_n) - f(x)| < \epsilon \frac{\delta^p}{2}$, completing the proof of the lemma.

Proof of Theorem 3.3.3

Let $f \in B^p(E,R)$ with bounded support, and $f \neq 0$. Without loss of generality we can assume that $f(x) = 0$ if $\|x\| \geq 1$, and $f(0) \neq 0$. We proceed to obtain a contradiction. Let A_n, α_n, h_n, $x_n \geq 0$ be inductively defined as follows, where A_n is a set, α_n a non-negative number, and h_n, x_n are in E. Let $A_0 = \{0\}$, $\alpha_0 = 0$, $h_0 = x_0 = 0$. Let A_{n+1} be the set of elements h in E such that

(1) $\|h\| < 1 - \|x_n\|$

(2) $|f(x_n + h) - f(x_n)| \leq \epsilon \dfrac{\|h\|^p}{2}$

(3) $\|x_n + h\|^p \geq \|x_n\|^p + \frac{1}{2}\|h\|^p$

(4) $\|x_n - x_k + h\|^p \leq \|x_n - x_k\|^p + 2\|h\|^p$

for $k = 0,1,2,\ldots,n$.

Define $\alpha_{n+1} = \sup\{\|h\| \mid h \in A_{n+1}\}$ and $x_{n+1} = x_n + h_{n+1}$, where h_{n+1} is chosen from A_{n+1} such that $\|h_{n+1}\| \geq \frac{1}{2}\alpha_{n+1}$. If $\|x_n\| < 1$, clearly A_{n+1} is not empty, since $0 \in A_{n+1}$. Further $\|x_{n+1}\| \leq \|x_n\| + \|h_{n+1}\| < 1$. From (3) $\sum_{i=1}^{n} \|h_i\|^p \leq 2\|x_n\|^p \leq 2$. Thus $\alpha_n \to 0$ as $n \to \infty$. It is verified from (4) that

$\sum_{m=0}^{i} \|h_{j+m}\| \le 2\sum_{m=0}^{i} \|h_{j+m}\|$. Hence $\{x_j\}$ is a Cauchy sequence. Let $x_j \to x$. It follows from (2) that $|f(x_n) - f(0)| \le \epsilon \sum_{j=1}^{n} \|h_j\|^p$ so that $|f(x) - f(0)| \le \epsilon$. We proceed to verify that $\|x\| = 1$, contradicting the choice of f.

Let $\|x\| < 1$. By Lemma 3.3.6 there is a $\delta > 0, 0 < \delta < \frac{1-\|x\|}{2}$, and a sequence $\{k_j\}$ in E such that $\|k_j\| = \delta$, $|f(x+k_j) - f(x)| \le \epsilon \frac{\delta^p}{8}$ and $\|z+k_j\|^p \to \|z\|^p + \delta^p$ for all $z \in E$. Pick a $\beta > 0$ such that $t^p - \frac{1}{10}\delta^p \le |t+s|^p \le t^p + \frac{1}{10}\delta^p$ if $0 \le t \le 2$, and $|s| \le \beta$. There is a N such that $\|x-x_N\| \le \mathrm{Min}(\frac{\delta}{4}, \beta)$, $|f(x) - f(x_N)| \le \epsilon\frac{\delta^p}{8}$, and $\alpha_{N+1} \le \frac{\delta}{2}$. There is a $k = k_i$, for sufficiently large i, such that $\|k\| = \delta$, and $|f(x+k) - f(x)| \le \epsilon\frac{\delta^p}{8}$, $\|x+k\|^p \ge \|x\|^p + \frac{3}{4}\delta^p$, and $\|x-x_j+k\|^p \le \|x-x_j\|^p + \frac{4}{3}\delta^p$ for $j = 0,1,2,\ldots,N$. Let $u = x - x_N + k$. Then since $\|u\| \ge \delta - \|x-x_N\|$, $u \notin A_{N+1}$, since $\alpha_{N+1} \le \frac{1}{2}\delta$.

We now prodeed to show $u \in A_{N+1}$, thus arriving at a contradiction, and completing the proof. By triangle inequality it follows that u satisfies condition (1). Since $\|x-x_N\| \le \beta$, $|f(x_N)| \le \epsilon\,\delta^p/8$ and $|f(x+k) - f(x)| \le \frac{\delta^p}{8}$, we have $\|u\|^p \ge \frac{3}{4}\delta^p$, and $|f(x_N+u) - f(x_N)| \le \epsilon\frac{\delta^p}{4} \le \frac{\epsilon}{2}\|u\|^p$ so that (2) is satisfied. Similarly from the definition of u, choice of x_N, β, k, and δ it is verified u satisfies conditions (3) and (4). Thus $u \in A_{N+1}$. The proof of the theorem is complete.

We conclude this section with a result on partitions of unity concerning B^p-smooth spaces. The proof is similar to that of the Theorem 3.2.13, and depends on methods developed in section 3.1. For detailed proofs, see [62] or [64].

3.3.7 <u>Theorem</u> [Torunczyk and Wells]

If E is a B^p-smooth Banach space and $\{U_\alpha\}$ is an open covering of E, there exists a locally finite partition of unity $\{f_\beta\}_{\beta\in X}$ subordinated to $\{U_\alpha\}$ such that $f_\beta \in B^p(E,R)$ for all $\beta \in X$.

3.4 Miscellaneous applications

We discuss in this section several specific applications of various results stated in section 3.1 to 3.3. Some of the applications are to arrive at certain negative results.

3.4.1 Theorem

If E is non-superreflexive Banach space, and F is a superreflexive Banach space, and $f:E \to F$ is a uniformly continuously differentiable function then for every bounded open set $U \subset E$, $f(\partial U)$ is dense in $f(U)$. Further if f_1, f_2 are two uniformly continuously differentiable functions on $E \to F$, coinciding on ∂U then $f_1 \equiv f_2$ on U.

Proof

Let $x \in U$. If possible let $f(x) \notin \overline{f(\partial U)}$, the closure of $f(\partial U)$. Then there is a ball U_ϵ with center at $f(x)$, and open set $G \supset \overline{f(\partial U)}$ such that $U_\epsilon \cap G = \emptyset$. Since F is U^1-smooth, Lemmas 3.2.3 and 3.2.5 assure that there is a uniformly continuously differentiable function ψ on F with support in U_ϵ, and $\psi(f(x)) = 1$. Let $g:E \to R$ be defined by $g(z) = \psi(f(z))$ if $z \in U$, and $g = 0$ on $E \sim U$. Since ψ vanishes in a neighborhood of $\overline{f(\partial U)}$, it is clear that g is a C^1-function. Now consider the halo $V = \{z | d(z,\overline{U}) < \Delta\}$ for some $\Delta > 0$ and the inequality

$$\|Dg(y) - Dg(z)\|_1 \leq \|D\psi(f(y))\|_2 \|Df(y) - Df(z)\|_3 + \\ + \|Df(z)\|_3 \|D\psi(f(y)) - D\psi(f(z))\|_2$$

where $\| \ \|_i$, $i = 1,2,3$ are respectively the norms in E^*, F^* and $L(E,F)$. It is verified that g is a uniformly continuously differentiable function with its support in the bounded set U. Thus E is U^1-smooth. Hence by the theorem 3.2.9 E is superreflexive leading to a contradiction.

The following two propositions show that certain smooth approximations are not possible on Banach spaces which are not superreflexive.

3.4.2 Proposition

Let E, F be two Banach spaces and E be non-superreflexive, and F be superreflexive. If $p:E \to F$ is a function such that $p(x) \to 0$ as $\|x\| \to \infty$, then there is no nontrivial C^2-function f with bounded support and with bounded second derivative on $E \to F$ such that $\|f(x)\|_1 \leq \|p(x)\|_1$, where $\|\cdot\|$, $\|\cdot\|_1$ are respectively the norms in the spaces E and F.

Proof

Let $f:E \to F$ be a nontrivial C^2-function such that $\|f(x)\|_1 \leq \|p(x)\|_1$. Let $x_0 \in E$ with $f(x_0) \neq 0$, and R be a positive number such that if $\|x-x_0\| \geq R$, then $\|p(x)\|_1 \leq \frac{1}{2}\|f(x_0)\|_1$. From the preceding theorem it follows that if $U = \{x \mid \|x-x_0\| < R\}$ then $f(\partial U)$ is dense in $f(U)$, but this contradicts the fact that $\|f(x)\|_1 \leq \|p(x)\|_1 \leq \frac{1}{2}\|f(x)\|_0$ for $x \in \partial U$, completing the proof.

3.4.3 Proposition

Let E, F be as in the proposition 3.4.2, and $p:E \to F$ be a bounded function with $p(x) \to 0$ as $\|x\| \to \infty$. If f is a C^2-function on $E \to F$, and f is not a C^3-function there does not exist a C^3-function g on E into F such that

(a) $\|f(x) - f(x)\|_1 \leq \|p(x)\|_1$, and

(b) $\|D^2f(x) - D^2g(x)\| \leq \|p(x)\|_1$

where $\|\cdot\|$ and $\|\cdot\|_1$ are the norms of E and F, and $\|\cdot\|_2$ is the norm of the space $L(E,L(E,F))$.

Proof

If possible let there be a C^3-function $g:E \to F$ satisfying inequalities (a) and (b). Since f is not of class C^3, $f \neq g$. Thus it may be assumed that $\|f(0) - g(0)\|_1 = a > 0$. From (a) and (b) it follows that $(f-g)$ is a nontrivial differentiable function with a Lipschitzian derivative. Since $p(x) \to 0$ as $\|x\| \to \infty$, the inequality $\|f(x) - g(x)\|_1 \leq \|p(x)\|_1$ contradicts the preceding proposition, completing the proof.

A well known theorem of Anderson and Kadec, see Bessaga and Pelczynski [6] asserts in part that all separable Banach spaces are homeomorphic. However the next corollary implies the nonexistence of homeomorphisms which are uniformly continuously differentiable between certain Banach spaces separable or not.

3.4.4 Proposition

If E, F are Banach spaces with F superreflexive and if there is a uniformly continuously differentiable homeomorphism on E into F, then E is also superreflexive.

Proof

Let $f:E \to F$ be a uniformly continuously differentiable homeomorphism. If possible let E be non-superreflexive. Hence E is not U^1-smooth. Consider the open balls $U_r = \{x \mid \|x\| < r\}$ in E. The hypothesis on f implies $f(\partial U_1)$, and $f(\partial U_2)$ are disjoint. However from Theorem 3.4.1 it follows that $f(0) \in f(\partial U_1) \cap f(\partial U_2)$ since f is a uniformly continuously differentiable homeomorphism, a contradiction completing the proof.

The preceding results 3.4.2 and 3.4.3 on smooth approximations have interesting applications in the context of differential equations in a Banach space. In the next example the essential part in one such application is indicated.

3.4.5 Example

Consider the problem of finding solutions F for the equation $D^2F(x) = \alpha(\|x\|)Q(x)$ where F is a real valued function on a Banach space E, with $F(x) \to 0$ as $\|x\| \to \infty$, α a nontrivial continuous function on $R \to R$, and Q is a continuous function on E into the Banach space of continuous symmetric bilinear forms of E, such that

$\alpha(\|x\|)Q(x) \neq 0$ for at least one point $x \in E$. Let E be a non-superreflexive space. If possible let the equation admit a solution.

Since $\sup_{x \in E} \|D^2F(x)\| < \infty$, and $f(x) \to 0$ as $\|x\| \to \infty$, it follows from Proposition 3.4.2 that F is identically 0 on E. Since $D^2F(x) \neq 0$ for at least one $x \in E$, the equation does not admit a solution if E is not superreflexive.

The next few applications are devoted to the structure of the Banach space ℓ_1. The results are consequences of Theorem 3.3.3.

3.4.6 <u>Proposition</u>

Let E be a Banach space which contains a subspace isomorphic with ℓ_1, and U be a bounded open subset of E. Suppose F is a Banach space with the norm differentiable away from 0. Let f be a continuous function on $\overline{U}$ into F which is differentiable in U. Then $f(\partial U)$ is dense in $f(\overline{U})$.

The proposition follows from Theorem 3.3.3 arguing as in the proof of the Theorem 3.4.1.

3.4.7 <u>Definition</u>

If F_1 and F_2 are two Banach spaces and $T: F_1 \to F_2$ is a continuous linear operator such that $\inf\{\|T_x\| : x \in f, \|x\| = 1\} = 0$ for every infinite dimensional subspace F of F_1, then T is said to be strictly singular, see Kato [31].

3.4.8 <u>Theorem</u>

Suppose E and F are Banach spaces and E contains a subspace isomorphic with ℓ_1. Suppose E is a bounded open set in E, $f: \overline{U} \to F$ is continuous and has a strictly singular derivative at each point in U. Then $f(\partial U)$ is dense in $f(\overline{U})$.

<u>Proof</u>

Without loss of generality we may suppose that $0 \in U$, diameter

$U \leq 1$ and $f(0) = 0$. It is enough to show that $0 \in$ closure of $f(\partial U)$. Let $T: \ell_1 \to E$ be an embedding, and $V = T^{-1}(U)$. V is an open bounded subset of ℓ_1, $T(\overline{V}) \subset \overline{U}$, and if $g = f \circ T$ is a continuous map on $\overline{V} \to F$, and has a strictly singular derivative at each point of V. We complete the proof by showing 0 is in the closure of $g(\partial V)$.

Suppose $\epsilon > 0$. Let Ω denote the set of all continuous maps $x: [0,\alpha] \to \ell_1$ (α depending on x) such that

(1) $x(0) = 0$, $0 \leq \alpha < \infty$, $\|x(s)-x(t)\| \leq |s-t|$ if $0 \leq s, t \leq \alpha$, and $x(\alpha) \in \overline{U}$

(2) $\| g(x(\alpha)) \| \leq \epsilon \frac{\alpha}{2}$

(3) $\| g(\alpha) \| \geq \frac{\alpha}{2}$.

Now $\| g(x(\alpha)) \| \leq \epsilon$ since $\alpha \leq 2 \|x(\alpha)\| \leq 2$. We complete the proof by showing that there is a path x such that $x(\alpha) \in U$.

Partially order Ω by the relation $x_1 \leq x_2$ is an extension on x_1. By applying Zorn's lemma it is verified that there is a maximal element $x_0: [0,\alpha_0] \to \ell_1$ in Ω. Now $x_0(\alpha_0) \in \partial U$. Suppose $x_0(\alpha_0) = y$, and $y \in U$. Since g is differentiable at y there is a $\delta > 0$ such that if $\|h\| \leq \delta$, $x + h \in U$ and $\|g(y+h)-g(y)-Dg(y)(h)\| \leq \epsilon \frac{\delta}{4}$. Let $E_n = \{y \in \ell_1 : y_1 = \ldots = y_n = 0\}$. There is a positive integer n such that if $h \in E_n$, and $\|h\| = \delta$ then $\|y+h\| \geq \|y\| + \frac{\delta}{2}$. Since $Dg(y): \ell_1 \to F$ is singular there is a $h_0 \in E_n$ of norm δ such that $\|Dg(y)(h_0)\| \leq \epsilon \frac{\delta}{4}$. We extend x_0 as follows. Define $x_1: [0,\alpha_0+\delta] \to \overline{U}$ by $x_1(t) = x_0(t)$ if $0 \leq t \leq \alpha_0$, and $x_1(t) = y + \frac{(t-\alpha_0)}{\delta} h_0$ if $\alpha_0 \leq t \leq \alpha_0+\delta_0$. It is verified that the path x_1 satisfies the conditions (1), (2), and (3) in the definition of Ω. Thus $x_1 \in \Omega$, and a contradiction is obtained. Thus $x_1(\alpha) \in \partial U$. This shows that $0 \in \overline{g(\partial V)}$, and the proof is complete.

The following corollaries are deduced from the preceding theorem noting that no infinite dimensional subspace of ℓ_1 has a separable dual, and every complety continuous linear mapping is strictly singular.

3.4.9 Corollary

If E contains a subspace isomorphic with ℓ_1 and F is a Banach space that is either reflexive or has a separable dual. Assume U is a bounded open set in E, and f is a continuous map on $\overline{U}$ into F, and differentiable in U. Then $f(\partial U)$ is dense in $f(\overline{U})$.

3.4.10 Corollary

Let E be a Banach space as in the preceding corollary, and F be an arbitrary Banach space. If U is a bounded open set, and f is a continuous function on $\overline{U}$ into F, and differentiable in U, then $f(\partial U)$ is dense in $f(\overline{U})$.

As a final application we mention a theorem of Kurzweil which is deduced from the results noted in 3.1.

If f is a Fréchet differentiable real valued function defined on the open unit ball of the space $C[0,1]$, then given $r_1 < r_2 < 1$ and $\epsilon > 0$, there exists at least one point x, $r_1 < \|x\| < r_2$, such that $|f(x) - f(0)| < \epsilon$.

Proof

We prove a stronger assertion namely if $0 < r < 1$ there is a x, $\|x\| = r$ such that $|f(x) - f(0)| < \epsilon$. Since density of $C[0,1] \neq$ density of $C[0,1]^*$, $C[0,1]$ does not admit a non-zero C^1-real valued function with bounded support. Thus if U is a bounded open subset of $C[0,1]$, f is continuous on $\overline{U}$, and differentiable in U, then $f(\partial U)$ is dense in $f(U)$, completing the proof of the theorem.

Chapter 4

Smoothness and Approximation in Banach Spaces

The purpose of this chapter is to discuss some approximation theorems for certain classes of functions defined on infinite dimensional Banach spaces. The motivation for the type of approximations studied here arises from the well known density theorems such as Weirstrass, Bernstein and Whitney which are widely used in analysis on finite dimensional spaces. To be more specific let $C(R^n)$ be the algebra of real valued continuous functions of class C^1 defined on R^n with the topology of uniform convergence of a function and its derivative on bounded subsets of R^n. A classical theorem of Bernstein asserts that the subalgebra $P(R^n)$ of all real polynomials on R^n is dense in $C(R^n)$. In the same spirit the approximation theorem of Whitney asserts that every continuous function on R^n is uniformly approximable by a real analytic function. This chapter is devoted to a discussion of the various extensions of the theorems of Bernstein and Whitney to infinite dimensional spaces. These generalizations and related approximation theorems have been extensively studied by several authors, Aron and Prolla [4] , Heble [24, 25] , Kurzweil [37] , Lesmes [40] , Llavona [41], Moulis [43] , Nachbin [44] , Nemirovski and Seminov [46] , Restrepo [54] and several others. Here we confine our attention to a few of these extensions which are tied up with the geometric structure, especially the approximation theoretic results discussed in [4, 37, 46, 54] . We discuss also the results on differential approximations of the type considered in [43] and [24, 25].

4.1 Polynomial algebras on a Banach space

We have already introduced the basic definitions related to a polynomial on a Banach space in 1.3.1. We need some more related concepts.

Let E, F be two Banach spaces, and N be the set of natural numbers. $P(^nE,F)$ is the Banach space of n-homogeneous continuous polynomials on E into F equipped with the norm defined by $\| P \| = \sup\{\| P_x \| \mid x \in E, \|x\| \leq 1\}$. As already noted in 1.3.1, we recall every such polynomial is a composition of a continuous symmetric n-linear transformation T, on $E \times E \ldots \times E$ into F, and the diagonal operator Δ_n on E into $E \times \ldots \times E$. We identify $P(^oE,F)$ with F. The subspace of $P(^nE,F)$ generated by monic n-homogeneous polynomials of the form $\phi^n \otimes y$, $n \in N$, $\phi \in E^*$, $y \in F$, where $(\phi^n \otimes y)(x) = \phi^n(x)y$ defined for each $x \in E$, is denoted by $P_f(^nE,F)$. The completion of the spaces $P_f(^nE,F)$ with respect to the norm $P(^nE,F)$ is denoted by $P_c(^nE,F)$ and is generally a proper subspace of $P(^nE,F)$. We denote the spaces of polynomials $\bigcup_{m\geq 0} \sum_{n\geq 0}^{m} P(^nE,F)$ and $\bigcup_{m\geq 0} \sum_{n\geq 0}^{m} P_f(^nE,F)$ respectively by $P(E,F)$ and $P_f(E,F)$.

The problems of approximation discussed here fall broadly into two classes. (1) Motivated by Stone-Weierstrass theorem in the finite dimensional setting it is natural to explore the possibility of approximating a uniformly continuous function on the unit ball (more generally on bounded subsets) of a given Banach space E uniformly by polynomials (more generally by differentiable functions).
(2) In the infinite dimensional spaces E, an equally natural problem is to discuss whether a function weakly continuous on bounded sets which is uniformly k-times differentiable on bounded sets is approximable uniformly by suitable polynomials in the topology of uniform convergence on bounded sets of a function f and its successive derivatives D^if, $1 \leq i \leq k$. The first problem in the case when E is a Hilbert space is discussed in [46], while the second type of approximation is discussed in [4], and [54].

We start with a counterexample, which establishes that in the infinite dimensional case the class of functions admitting uniform approximation by polynomials on bounded sets is very small, thus disposing in part the first problem raised in the beginning of the preceding paragraph. This example is discussed in [46].

Let V be a closed ball of a separable Hilbert space, and $C^{\infty}(V)$ be the algebra of uniform limits of polynomials on the ball V, and let $A_{\infty}(V)$ be the algebra of uniform limits of functions in $D_b^{\infty}(V)$, the space of functions defined in a neighborhood of V, and which are infinitely boundedly differentiable on V.

4.1.1 <u>Example</u>

$C_{\infty}(V) \neq A_{\infty}(V)$.

<u>Proof</u>

Let R^n be the n-dimensional Euclidean space and V_n its unit balls. In V_n choose a maximal set Γ_n of points such that any two points in Γ_n are at a distance 1/2 apart. Then comparing the Lebesgue measures of V_n, and the balls of radius 1/2 centered at points in Γ_n, it follows that $R(n) = \text{Card}(\Gamma_n) \geq 2^n$.

Let P_n^s be the space of polynomials on R^n of degree not exceeding s. It is verified that $\dim P_n^s \leq sn^s + 1$. Let $n(s)$ be chosen so large that $2^{n(s)} > s\, n(s)^{\ s} + 1$. Let T_n be the space of functions on Γ_n with uniform norm, and let $\tau_s : P^s_{n(s)} \to T_{n(s)}$ be the linear mapping carrying a function into its restriction to $\Gamma_{n(s)}$. Since $\dim T_{n(s)} = R(n(s)) > \dim P^s_{n(s)}$ the range of τ_s is a proper subspace of $T_{n(s)}$; therefore there exists a $\phi_s \in T_{n(s)}$ such that $\|\phi_s\| = 1$ and $\|\phi_s - \tau_s p\| \geq 1$ for all $p \in P^s_{n(s)}$.

Now let $V_{0,4}$ be the ball of radius 4 center at 0, in H, and let $\{W_i\}_{i\geq 1}$ be balls of radius 1 contained in $V_{0,4}$, with distance between centers at least 5/2. We may assume that $V_{n(s)}$ is the section of W_s by a finite dimensional plane L_s of dimension $n(s)$ passing through its center so that $\Gamma_{n(s)} \subset W_s$. Let $\Gamma = \bigcup_{i=1}^{\infty} \Gamma_{n(s)}$, and let ϕ be the function on Γ defined by ϕ_s = the restriction of ϕ to $\Gamma_{n(s)}$, $\phi/\Gamma_{n(s)}$.

Note that $\|\phi\| = 1$, and for any polynomial p in H we do have

$\| p - \phi \|_\Gamma \geq 1$. In fact if $\deg p = s$, then

$$\| p - \phi \|_\Gamma \geq \| p - \phi_s \|_{\Gamma(n(s))}$$
$$= \| \tau_s(p/L_s) - \phi_s \|_{\Gamma(n(s))} \geq 1.$$

On the other hand, if $\Gamma = \{x_i\}_{i \geq 1}$, then the points x_i and x_j are at distance at least 1/2 apart $(i \neq j)$. If $\theta(t)$ is a smooth function on the line, with $\theta(0) = 1$ and $\theta(t) = 0$ for $|t| > 1/8$, then the function

$$\psi(x) = \sum_{i \geq 1} \phi(x_i)\theta(\| x - x_i \|^2)$$

lies in $D_b^\infty(H)$ and $\psi/\Gamma = \phi$. Therefore for any polynomial p in H we have

$$\| p - \psi \| V_{0,4} \geq 1 \quad \text{and} \quad \psi \notin C_\infty(V_{0,4}).$$

Thus $C^\infty(V) \neq A^\infty(V)$, completing the proof.

4.2 Approximation by smooth functions

In contrast to the preceding example it is possible to establish uniform approximation theorems on bounded sets by smooth functions in the case of certain Banach spaces. More precisely we have the following result, stated and proved in [46].

4.2.1 Theorem

Every uniformly continuous function on a ball in a separable Hilbert space H is the uniform limit of restrictions of functions which are uniformly continuously differentiable on bounded subsets, in fact by C^1 functions having derivatives satisfying a Lipschitz condition.

Before proceeding to the proof, we denote the set of all functions which are uniformly continuously differentiable on bounded subsets of H, by $D_u'(H)$. Further H is a separable infinite dimensional Hilbert space and $V_{0,r}$ is the closed ball of radius r with center at 0 in H.

Before presenting the proof of the Theorem 4.2.1, we recall some useful results from the theory of metric approximation in Banach spaces, Singer [55].

4.2.2 <u>Definition</u>

If A is a subset of a Banach space E, and $x \in E$, then the distance of x from A, $d(x,A)$ is defined by

$$d(x,A) = \inf \{ \| x-y \| \,\big|\, y \in A\}.$$

4.2.3 <u>Lemma</u>

If A is a closed convex subset of a reflexive Banach space E, then for each $x \in E$, there exists a $p(x) \in A$ such that $d(x,A) = \|x-p(x)\|$. Further if the norm is strictly convex then $p(x)$ is unique.

<u>Proof</u>

If $x \in E$, consider the closed ball B center x, radius $d(x,A) + \delta$, where δ is a positive number. Now $B \cap A$ is a bounded closed convex set. Since E is reflexive, $B \cap A$ is weakly compact. Since the norm is weakly lower semi-continuous, there exists a point $p(x) \in A$, with the desired property. Further if E is strictly convex let, if possible, $y,z \in A$ such that

$$\|x-y\| = \|x-z\| = d(x,A).$$

Since A is convex, $\frac{1}{2}(y+z) \in A$, and

$$d(x,A) \leq \|x - \tfrac{1}{2}(y+z)\| \leq \tfrac{1}{2} \quad \|x-y\| + \|x-z\| \quad = d(x,A).$$

Hence $\|x-y\| = \|x-z\| = \|x - \frac{1}{2}(y+z)\|$ and the strict convexity of the norm implies $x = y = x - z$. Thus $y = z$, completing the proof of the proposition.

4.2.4 <u>Remark</u>

If E is uniformly convex, and A is a closed convex set then the map $x \to p(x)$ on $E \to A$ is continuous, where $p(x)$ is as described in 4.2.3.

For a proof of the above remark, see appendices 1 and 2 in Singer [55].

The Proposition 4.2.5 below is well-known, see Restrepo [53] or Phelps [50].

4.2.5 Proposition

Let E be a Banach space with norm differentiable away from 0. Suppose for every $x \in E$, and some $p(x) \in A$, $\|x-p(x)\| = d(x,A)$. Then the function $g: E \sim A \to R$ defined by $g(x) = d(x,A)$ is differentiable and $Dg(x) = D\|\ \|(x-p(x))$, where $D\|\ \|(x)$ is the derivative of the norm at x.

Proof

For $x \in A$

$$\|x+y - p(x)\| = \|x-p(x)\| + D\|\ \|(x-p(x))y + 0(\|y\|),$$

for any y with $p(x) + y \in A$,

$$\|x - (p(x) + y)\| \geq \|x-p(x)\|$$

which implies $D\|\ \|(x-p(x))(y) \leq 0$. Thus the hyperplane

$$L = \{y | D\|\ \|(x-p(x))(y-p(x)) = 0\}$$

is a support hyperplane for A at $p(x)$, and $d(x+y,L) \leq (d(x+y,A) \leq$ $\leq (d(x+y,p(x))$, so that

$$\|x-p(x)\| + D\|\ \|(x-p(x))(y) \leq d(x+y,A) \leq \|x-p(x)\| + D\|\ \|(x-p(x))[y] + 0(\|y\|)).$$

Thus $0 \leq d(x+y,A) - d(x,A) - D\|\ \|(x-p(x))(y) \leq o(\|y\|)$.

Thus $g(x)$ is differentiable, and $Dg(x) = D\|\ \|(x-p(x))$.

4.2.6 Definition

If U is an open subset of a Banach space and M is a positive number, then the set of real valued C^1-functions f on U such that $\|Df(x) - Df(y)\| \leq M\|x-y\|$, is noted as $B^1_M(U)$.

4.2.7 Remark

For example by direct computation of derivatives one can verify that if E is a Hilbert space, and $U = \{x \mid \|x\| > \alpha\}$, then the norm $\|\ \|$ of E is in $B^1_M(U)$, where $M = 2/\alpha$.

4.2.8 Proposition

If A is a closed convex subset of a Banach space E, and $\| \ \| \in B_M^1\{x \big| \|x\| > \alpha\}$ for some positive number α, then $d(x,A) \in B_M^1\{x|d(x,A) > \alpha\}$, if for each x in E there is a nearest point $p(x)$ in A.

Since the proof is similar to the proof of Proposiiton 4.2.5 it is not supplied here.

Since the $\| \ \|$ of a Hilbert space is in $B_M^1\{x \big| \|x\| > \alpha\}$, the following corollary is deduced from Proposition 4.2.8.

4.2.9 Corollary

If $V_{0,r}$ is the closed ball center 0 and radius r in a Hilbert space H, and K is a closed convex subset of H, then the function $\alpha(x,K_r)$ where $K_r = K + V_{0,r}$ is differentiable outside any set K_{r_0}, with $r_0 > r$, with the derivative satisfying a Lipschitz condition.

We present first the proof of the Theorem 4.2.1 when f is a convex function.

4.2.10 Theorem

Let f be a uniformly continuous convex unimodal function on the ball V. Then for any $\epsilon > 0$ there exists a $g \in D_u'(H)$ with derivative satisfying a Lipschitz condition, such that $\|f - g_V\| < \epsilon$, where $g_V = g|V$.

Proof

Let $|f(x)| \leq 1$ on V. Subdivide the interval $[-1,1]$ into subintervals Δ_i, $|i| \leq M$, with disjoint interiors each of length $\leq \epsilon/2$ and let m_i be the midpoint of Δ_i. Choose $\delta > 0$ so small that $d(m_i, \Delta_j) > \delta$, if $i \neq j$. Let

$$Q_i = f^{-1}(\Delta_i)$$

and

$$S_i = \{x \in V | f(x) \leq m_i\}.$$

The sets S_i are convex and closed. Choose $\alpha > 0$ so small that for any $x, y \in V$ satisfying $\|x-y\| \leq \alpha$, we have $|f(x) - f(y)| < \delta$. Let $\tilde{S}_i = S_i + V_{0,\alpha/2}$. We now construct functions $\tilde{\phi}_i$ on H such that (1) $\tilde{\phi}_i(x) \geq 0$ on H; (2) $\sum_{i=-M}^{M} \tilde{\phi}_i(x) \geq 1$ on V and (3) if $\tilde{\phi}_i(x) \neq 0$, then $|f(x) - m_i| < \epsilon$ for $x \in V$; (4) $\tilde{\phi}_i \in D_u^1(H)$, and the derivative of each $\tilde{\phi}_i$ satisfies a Lipschitz condition.

The construction is carried as follows. We note that $Q_i \subset S_{i+1}$ and $P(Q_i, \tilde{S}_{i-1}) \geq \alpha/4$, for $|i| \leq M - 1$ which follow from the definitions of the sets Q_i and $\tilde{S}_i$.

Now applying the lemma to each of the sets S_i, it is verified that the functions $r_i(x) = P(x, \tilde{S}_i)$ are continuously differentiable outside $\tilde{S}_i$ and have a derivative satisfying a Lipschitz condition outside any r-neighborhood of the sets $\tilde{S}_i$. Let $\theta(t)$ be a C^∞-function on $R \to R$, such that $\theta(t) = 0$ if $|t| \leq \alpha/8$, and $\theta(t) = 1$ if $|t| \geq \alpha/4$, with $0 \leq \theta(t) \leq 1$, then the functions $\tilde{r}_i(x) = \theta(r_i(x))$ satisfy condition (4).

Now let $\tilde{\phi}_{-M}(x) = 1 - \tilde{r}_{-M}(x)$, $\tilde{\phi}_M(x) = \tilde{r}_{M-1}(x)$ and $\tilde{\phi}_i(x) = \tilde{r}_{i-1}(x)(1 - \tilde{r}_i(x))$ for $|i| \leq M - 1$. The functions $\{\tilde{\phi}_i\}_{|i| \leq M-1}$ verify the conditions (1) to (4).

Since $\sum_{-M}^{M} \tilde{\phi}_i$ is uniformly continuous on H, and is ≥ 1 for all x in V it is possible to find a nonnegative function $\tilde{\phi} = 0$ on V satisfying the condition (4), and such that

$$\sum_{-M}^{M} [\tilde{\phi}_i(x) + \tilde{\phi}(x)] \geq 1 \quad \text{on } H.$$

Now let

$$\phi_i(x) = \frac{\tilde{\phi}_i(x)}{\tilde{\phi}(x) + \sum_{-M}^{M} \tilde{\phi}_i(x)}, \quad |i| \leq M.$$

Clearly $\{\phi_i\}$ satisfy condition (4). Let $g(x) = \sum_{i=-M}^{M} M_i \tilde{\phi}_i(x)$. Then it is verified that $\sup_{x \in V} |f(x) - g(x)| \leq \epsilon$.

To pass on from convex f to the case of arbitrary functions f, we note the following result.

4.2.11 Theorem

The polynomials in functions convex and uniformly continuous on a closed ball V are dense in the space of uniformly continuous functions on V with supremum norm.

Proof

Let $\mathcal{A}$ be the smallest uniformly closed subalgebra of functions on V containing all convex functions which are uniformly continuous on B. We verify that $\mathcal{A}$ is same as the set of all functions which are uniformly continuous on V. The proof of this claim is completed in 2 steps.

1. Let $0 \leq r \leq 1$, and $S_{0,r}$ be the boundary of $V_{0,r}$. Then the restrictions of functions in $\mathcal{A}$ to $S_{0,r}$ are dense in the space of uniformly continuous functions on $S_{0,r}$ in the supremum norm. To verify this claim it is sufficient to prove that for each closed set $Q \subset S_{0,r}$ and $\epsilon > 0$ there exists a function $g \in \mathcal{A}$ such that $g|Q = 1$, and $g(x) = 0$ for all $x \in S_{0,r}$ such that $d(x,Q) \geq \epsilon$. Once this is done we can proceed as in the proof of the preceding theorem to verify the assertion in the beginning of this paragraph. We proceed to construct g. Let $\hat{Q}$ be the closed convex hull of Q. Then the function $d(x,\hat{Q})$ is uniformly continuous and convex in V so that $d(\cdot,\hat{Q}) \in \mathcal{A}$. Since the norm in a Hilbert space is uniformly convex, for some $N = N(r)$ and all $x \in S_{0,r}$ we have $N(r)\ d(x,\hat{Q})^{1/2} \geq d(x,\hat{Q}) \geq d(x,\hat{Q})$. From this inequality it follows that if $\theta(t)$ is a function on $R \to R$, $\theta(0) = 1$, $\theta(t) = 0$ outside a sufficiently small neighborhood of the origin, then the function g, where $g(x) = \theta(d(x,\hat{Q}))$ is the desired one.

2. To complete the proof of Theorem 4.2.1, let $\epsilon > 0$ and f be a function uniformly continuously differentiable on V. We proceed to

find g such that $\sup_{x \in V} |f(x) - g(x)| \leq \epsilon$. From the step 1, for each r, $0 \leq r \leq 1$, there is a function $g_r \in \mathcal{A}$, such that $\sup_{x \in S_{0,r}} |f(x) - g_r(x)| \leq \epsilon/2$. Since f, g_r are uniformly continuous, it follows there exists an interval $\Delta_r \ni r$, such that for all $x \in V$ with $\|x\| \in \Delta_r$ we have $|f(x) - g_r(x)| \leq \epsilon$. Let $\{\Delta_{r_j}\}_{j=1}^N$ be a finite subcovering of $[0,1]$, extracted from $\{\Delta_r\}_{0 \leq r \leq 1}$, and let $\{\theta_i\}_{i=1}^N$ be partition of unity subordinated to $\{\Delta_{r_j}\}_{j=1}^N$ on $[0,1]$. Now let

$$g(x) = \sum_{i=1}^{N} g_{r_i}(x)\, \theta_i(\| x \|).$$

It is verified that $g(x)$ is the desired function, completing the proof of the theorem.

From the preceding two theorems the theorem 4.2.1 follows at once.

4.2.12 Remark

Since neither separability nor Euclidean structure of a Hilbert space, except uniform convexity and uniform smoothness of H, are used in the proofs of the preceding two theorems, we note that if E is a Banach space whose norm is uniformly convex and uniformly smooth then the restrictions of functions which are uniformly continuously differentiable in E to the ball V are dense in the space of functions which are uniformly continuous on V, in the supremum norm.

4.3 Extension of Bernstein's theorem to infinite dimensional Banach spaces

In this section we discuss the second problem raised in §4.1. Here we are given a weakly continuous function f on a Banach space E which is k-times uniformly differentiable on bounded sets. Now given $\epsilon > 0$, the problem is to find a polynomial p_ϵ on E so that

$$\sup_{\substack{x \in M \\ 1 \leq i \leq k}} \{ |f(x) - p\ (x)|,\ \|D^i f(x) - D^i p_\epsilon\ (x)\| \} < \epsilon .$$

where M is an arbitrary bounded subset of E. This has been studied by Restrepo [54], and Aron and Prolla [4], in the case when E is reflexive space with property B, and when E is an arbitrary Banach space with E^* admitting the bounded approximation property respectively. The results presented here are taken from [4].

Before proceeding to present the results we recall a few definitions and certain basic results concerning polynomials, and relevant geometric properties of Banach spaces.

If E is a Banach space let us denote by E^* as usual the dual of E. Let us denote the n-fold cartesian product of E^* by $\overset{n}{\underset{i=1}{X}} E^*$, for $n \geq 1$. For each n-sequence of positive integers $(\alpha_1, \alpha_2, \ldots, \alpha_n)$ and for each $u = (u_1, u_2, \ldots, u_n) \in \overset{n}{\underset{i=1}{X}} E^*$, let us denote by $u^\alpha(x) = \prod_{i=1}^{n} u_i^{\alpha_i}(x)$, where $\alpha = \sum_{i=1}^{n} |\alpha_i|$. Then it is verified that the set of polynomials of the form $\sum_{|\alpha| \leq m} a_\alpha u^\alpha$, is the same as $P_f(E,R)$ introduced earlier in Section 4.1, if a_α's are real numbers.

4.3.1 Definition

Following the terminology in [4], we define $P_W(^mE,F)$ as the subspace of $P(^mE,F)$ of F-valued m-homogeneous polynomials on E which are weakly uniformly continuous on bounded subsets of E. Further $C_W^m(E,F)$ is the space of m-times continuously differentiable functions $f:E \to F$ satisfying the conditions (a) $D^j f(x) \in P_W(^jE,F)$, $j \leq m$, and (b) $D^j f: E \to P_W(^jE,F)$ is weakly uniformly continuous on bounded subsets of $E (j \leq m)$. Further $C_W^\infty(E,F)$, by definition, $= \bigcap_{m=0}^{\infty} C_W^m(E,F)$.

We topologize the space $C_W^m(E,F)$ by the locally convex topology defined by the seminorms

$$f \in C_W^m(E,F) \to \sup\{\| D^j f(x) \| ;\ x \in B,\ j \leq m\}$$

where B is allowed to range over the bounded subsets of E. This topology is denoted here by t_b^m. In passing we note that $(C_W^m(E,F); t_b^m)$ is complete if E,F are Banach spaces, for all m, $1 \leq m < \infty$.

It follows easily that every polynomial $u^{\alpha}(x)$ (*-polynomials) on E is in $C_W^m(E,R)$ for all values of m, and $P_f(E,F) \subset C_W^m(E,F)$.

We proceed to collect some properties of the spaces $C_W^m(E,F)$, and the mappings on E which are weakly continuous on bounded subsets.

We state two lemmas without proof which follow from definitions.

4.3.2 Lemma

Let E be a Banach space, and p be in $P_f(E,R)$. Then (a) p is weakly continuous, (b) the derivative $p':E \to E^*$ is weakly continuous, and (c) $p'(E)$ is finite dimensional.

4.3.3 Lemma

If E,F are two Banach spaces and $f:E \to F$ continuous and $P:E \to F$ is a continuous linear operator such that P(E) is finite dimensional then (a) $f \circ P$ is weakly continuous (b) $(f \circ P)' = P^* \circ f' \circ P$, where P^* is the adjoint of P.

The next result characterizes functions on a reflexive Banace space E to R which are weakly continuous on bounded sets in E.

4.3.4 Theorem

Let E be a reflexive Banach space E. Then $f:E \to R$ is weakly continuous on bounded sets iff there is a sequence of *-polynomials p_n converging to f uniformly on bounded sets.

Proof

Since every closed ball is weakly compact in a reflexive space E, f is weakly continuous if $p_n \to f$ uniformly on bounded sets, since each p_n is weakly continuous as noted in the paragraph following definition 4.3.1. The converse is an easy consequence of Stone-Weierstrass theorem thus completing the proof.

The following propositions clarify in part the relation between $C^m(E,F)$ and $C_W^m(E,F)$. The first proposition follows from the definitions.

4.3.5 Proposition

Let $f \in C^m(E,F)$ satisfy the following conditions:

(a) $f \in C_W^{m-1}(E,F)$;

(b) f is uniformly differentiable of order m on bounded sets;

(c) $D^m f(x) \in P_W({}^m E,F)$.

Then $f \in C_W^m(E,F)$.

4.3.6 Proposition

If $f \in C_W^m(E,F)$, then f is uniformly differentiable of order j, for all $j \leq m$.

Proof

It suffices to take the bounded set in the definition of $C_W^m(E,F)$ as the closed unit ball B. Let $\epsilon > 0$. The proof is completed by exhibiting a $\delta > 0$ and $\{\phi_i\}_{i=1} \subset E^*$ so that if $x,y \in B$, $|\phi_i(x-y)| < \delta$ $(i = 1,2,...,n)$ then

$$\left\| \frac{D^m f(x)}{m!} - \frac{D^m f(y)}{m!} \right\| < \epsilon .$$

Choose $\delta_1 > 0$ such that if $x \in B$, $h \in E$ with $\|h\| \leq \delta_1$, then

$$\| f(x+h) - \sum \frac{D^i f(x)}{i!} (h^m) \| \leq \epsilon \|h\|^m.$$

Since $f \in C_W^{m-1}(E,F)$, there exist $\{\phi_i\}_{i=1}^n \subset E^*$, such that if $x,y \in (1 + \delta_1)B$, $|\phi_i(x - y)| < \delta$, $1 \leq i \leq n$, then

$$\left\| \frac{D^j f(x)}{j!} - \frac{D^j f(y)}{j!} \right\| \leq \frac{\epsilon \delta_1^{m-j}}{m+1} , \quad (j \leq m - 1).$$

Thus for $x,y \in B$, $h \in E$, $\|h\| = 1$

$$\left\| \frac{D^m f(x)(h^m)}{m!} - \frac{D^m f(y)(h^m)}{m!} \right\|$$

$$= \frac{1}{\delta_1^m} \left\| \frac{D^m f(x)}{m!} ((\delta_1 h)^m) - \frac{D^m f(y)}{m!} ((\delta_1 h)^m) \right\|$$

$$\leq \frac{1}{\delta_1^m} \left\| \frac{D^m f(x)}{m!} ((\delta_1 h)^m) - (f(x+\delta_1 h) - f(x_1) - \sum_{j=1}^{m-1} \frac{Df(x)}{j!} ((\delta_1 h)^j) \right\|$$

$$+ \frac{1}{\delta_1^m} \left\| \frac{D^m f(y)}{m!} ((\delta_1 h)^m) - (f(y+\delta_1 h) - f(y) - \sum_{j=1}^{m-1} \frac{Df(y)}{j!} ((\delta_1 h)^j) \right\|$$

$$+ \frac{1}{\delta_1^m} [\| f(y+\delta_1 h) - f(x+\delta_1 h) \| + \sum_{j=0}^{m-1} \left\| \frac{D^j f(y)}{j!} ((\delta_1 h)^j) - \frac{D^j f(x)}{j!} ((\delta_1 h)^j) \right\|]$$

$\leq 3\epsilon$, completing the proof.

The following result follows from the definitions.

4.3.7 Proposition

If $f \in C_W^m(E,F)$, then f is uniformly differentiable of order j, $j \leq m$.

As a corollary of the preceding proposition it follows that $C_W^m(E,F)$ is the same as the set of functions f in $C^m(E,F)$ such that

(a) $f: E \to F$ is weakly continuous on bounded sets.

(b) $D^j f(x) \in P_W(^j E,F)$, and

(c) f is uniformly differentiable of order j $(j \leq m)$.

Definition

A Banach space E is said to have the approximation property (A.P.), if for every compact set K in E, for every $\epsilon > 0$, there exists an operator $T: E \to F$ of finite rank such that $\| Tx - x \| < \epsilon$ for all $x \in K$. Clearly T depends on ϵ and K. If it is possible to choose T such that for every $\epsilon > 0$, and for all K, $\|T\| \leq \lambda$, λ a constant independent of ϵ and K, then E is said to have the bounded approximation property (B.A.P.).

We state a theorem which establishes the relation between A.P. and the spaces of polynomials. We sketch a proof of the same.

4.3.8 Theorem

E^* has A.P. iff for all Banach spaces F and for all positive integers m,

$$P_W(^m E,F) = P_C(^m E,F).$$

Before proceeding to the proof of the theorem, we state three useful lemmas.

4.3.9 Lemma

Let E, F be two real Banach spaces, and let $f:E \to F$ be weakly uniformly continuous on bounded sets. Let B be a bounded subset of E. Then $f(B)$ is precompact.

Proof

For $\epsilon > 0$ and $x \in B$, let $V(x, \epsilon)$ be the set $\{t \in F \mid \|t - f(x)\| < \epsilon\}$. Since f is weakly uniformly continuous on bounded sets, there are $\delta > 0$, $\{\phi_i\}_{i=1}^k \subset E^*$, such that whenever $x, y \in B$, and $|\phi_i(x-y)| < \delta$ $(i = 1,2,\ldots,k)$, then $\|f(x) - f(y)\| < \epsilon$. The mapping $\phi:E \to R^k$ defined by $\phi(x) = (\phi_1(x), \phi_2(x), \ldots, \phi_k(x))$ is continuous. Hence $\phi(B)$ is a precompact subset of R^k. Thus if $x \in B$, then there is a set $\{x_i\}_{i=1}^n \subset B$ such that for some x_j, $|\phi_i(x) - {}_i(x_j)| < \delta$, $i = 1,\ldots,k$. Thus $\|f(x) - f(x_j)\| < \epsilon$. Hence

$$f(B) \subset \bigcup_{i=1}^{n} V(x_i, \epsilon),$$

verifying that $f(B)$ is precompact in F.

4.3.10 Lemma

If $A:E \to F$ is a continuous linear mapping, A is compact if and only if A is weakly (uniformly) continuous on bounded subsets of E.

Proof

If B is the unit ball of A, let $X = \overline{A(B)} + \overline{A(B)}$. Since A is compact, X is a compact subset of F. Now for each $x \in X$ pick $\psi_x \in F^*$ such that $\psi_x(x) = \|x\|$, with $\|\psi_x\| = 1$. Since the function $x \in E \to \|x\| - |\psi_x(x)|$ is a continuous real valued function there is a neighborhood U_x of x such that $\|y\| \leq |\psi_x(x)| + \epsilon$ for all $y \in U_x$ for a given $\epsilon > 0$. Since X is compact we can find $\{\psi_i\}_{i=1}^n \subset F^*$ such that for all $x \in X$, $\|x\| \leq \sup_{1 \leq i \leq n} |\psi_i(x)| + \epsilon$.

Now noting $\psi_i \circ A \in E^*$, it follows that if $x,y \in B$, satisfy $|\psi_i \circ A(x-y)| < \epsilon$, $1 \leq i \leq n$, then $\|A(x-y)\| < \epsilon$.

The converse is contained in the preceding lemma.

4.3.11 Lemma

Let E,F be as above, and $\Delta_m : E \to \prod_1^m E$ be the diagonal map and A be a continuous m-linear form on E into F. Let $C:E \to P({}^{m-1}E,F)$ be defined by $C(x)(z) = A(x,z,z,\ldots,z)$ for x,z in E. Then C is a compact linear mapping with range in $P_W({}^{m-1}E,F)$.

This lemma follows from lemma 4.3.9 after using the polarization identity [2].

Proof of Theorem 4.3.8

Let E^* have the A.P. Since $P_W({}^mE,F) \supset P_f({}^mE,F)$ and complete $P_W({}^mE,F) \supset P_C({}^mE,F)$. Thus we need only show the reverse inclusion. Let $m = 1$. If $A:E \to F$ be weakly uniformly continuous on bounded sets, then it is compact by lemma 4.3.9. Since E^* has A.P. A is the limit of operators in $E^* \otimes F$, so that $A \in P_C({}^1E,F)$. Assuming now that the result is true for $i = 1,2,\ldots,m-1$, and $P \in P_W({}^mE,F)$, by the preceding lemma the mapping $C:E \to P_W({}^{m-1}E,F)$ is compact and linear. Since E^* has A.P. given $\epsilon > 0$, there is a finite rank operator $\sum_{i=1}^{k} \phi_i \otimes P_i$, $\phi_i \in E^*$, $P_i \in P_W({}^{m-1}E,F)$ such that $\| C - \sum_{i=1}^{k} \phi_i P_i \| < \epsilon$. Since $P(x) = C(x)(x)$, it follows that $\| C - \sum_{i=1}^{k} \phi_i(x) P_i(x) \| < \epsilon$. Now by induction hypothesis, it is possible to pick $Q_i \in P_f({}^{m-1}E,F)$, such that $\|P_i(x) - Q_i(x)\| < \frac{\epsilon}{k\|\phi_i\|}$ $(i = 1,2,\ldots,k)$. Thus $\|P - \sum_{i=1}^{k} \phi_i \otimes Q_i\| < 2\epsilon$, proving that $P \in P_C({}^mE,F)$ as desired.

To complete the proof of the theorem note that if $P_W({}^1E,F) = P_C({}^1E,F)$ for all Banach spaces F, then by a well known thoerem of Grothendieck, it follows that E^* has the A.P.

4.3.12 Definition

A polynomial algebra $A \subset C^m(U,F)$ is a Nachbin polynomial algebra [3], if (i) for each $x \in U$, there is a function $g \in A$ such that $g(x) \neq 0$,

(ii) for every pair $x,y \in U$, $x \neq y$, there is a $g \in A$ such that $g(x) \neq g(y)$,

(iii) for every $(x,u) \in U \times E$, with $u \neq 0$, there is a $g \in A$ such that $Dg(x)(u) \neq 0$.

It is proved in Nachbin [44] that if E is finite dimensional then an algebra $A \subset C^m(U,R)$ is dense in $C^m(U,R)$ iff A satisfies the conditions (i)-(iii) stated above.

We now proceed to prove the generalization of Bernstein's theorem to the infinite dimensional case stated and proved in [4].

We sketch the proof of the generalization when $m = 0$, [4]. In the proof we denote the space $C^0_W(E,F)$ by X.

4.3.13 Theorem

For arbitrary Banach spaces E and F the τ^0_b completion of $P_f(E,F)$ is $C^0_W(E,F)$.

Proof

It is verified that (X, τ^0_b) is complete. Thus it is enough to verify that $P_f(E,F)$ is dense in the space X. Let $f \in X$, $B \subset E$ be bounded, and $\epsilon > 0$. Let $\{\phi_i\}_{i=1}^n \subset E^*$, $\delta > 0$ be so chosen that if $|\phi_i(x,y)| < \delta$, $1 \leq i \leq n$ for $x,y \in B$, then $|f(x) - f(y)| < \epsilon$. Define $\phi: E \to R^n$ by $\phi(x) = (\phi_i(x))_{i=1}^n$. Choose $\{x_i\}_{i=1}^m \subset B$ such that $\|\phi(x) - \phi(x_i)\| < \frac{\delta}{2}$ for some i, $1 \leq i \leq n$, where R^n is given the supremum norm. Consider the balls $B(\phi(x_i), \delta/2) = B_i$, and let $Y = \bigcup_{i=1}^m B_i$. Choose continuous nonnegative real valued functions $\{h_i\}_{i=1}^m$ on R^n into R such that $\sum_{i=1}^m h_i(y) \leq 1$, for all $y \in R^n$, and $\sum_{i=1}^m h_i(y) = 1$ for $y \in Y$, with support $h_i \subset B_i$, $1 \leq i \leq m$.

Choose polynomials q_i on $R^n \to R$ such that for all $y \in Y$

$$|q_i(y) - h_i(y)| < \frac{\epsilon}{m \operatorname{Max}(1, \|f(x_i)\|)} .$$

Let $g:E \to F$ be defined by $g(x) = \sum_{i=1}^{m} q_i(\phi(x)) \cdot f(x_i)$. Then $g \in P_f(E,F)$, and for all $x \in B$, it is verified that $\|g(x)-f(x)\| < 2\epsilon$, by using the triangle inequality together with the fact that $\sum_{i=1}^{m} h_i(\phi(x)) = 1$ for $x \in B$. This completes the proof.

Before proving the main approximation theorem we state a few lemmas.

4.3.14 Lemma

Let E be a real Banach space and K be a precompact subset of $C_W^m(E,F)$. If B is a bounded subset of E, and $i \leq m$, then

$$L_i = \{D^i f(x) \mid x \in B, \ f \in K\}$$

is a precompact subset of $P_w({}^iE,F)$.

The lemma follows from the fact that if $T:E \to F$ is weakly uniformly continuous, then $T(B)$ is precompact subset of F, if B is a bounded set in E.

From the preceding lemma, proposition and the definition of the space $C_W^m(E,F)$, the following lemma is deduced.

4.3.15 Lemma

Let E be a real Banach space such that E^* has B.A.P. with constant C. Let K be a precompact subset of $C_W^m(E,F)$. Given $j \in N$, $j \leq m$, $\epsilon > 0$, and $B \subset E$ a bounded set, there exists a continuous linear mapping $\Pi:E \to E$, $\|\Pi\| \leq C$, such that $\|D^i(f(\Pi x)) - D^i f(x)\| \leq \epsilon$, if $x \in B$, $f \in K$, and $i \leq j$.

The next theorem is the main theorem in this section from which the generalized Bernstein's theorem is deduced as a corollary.

4.3.16 Theorem

Let E,F be two Banach spaces with E^* verifying the B.A.P. with constant C, and let m be a positive integer. A polynomial algebra $A \subset X$ $(= C_w^m(E,F))$ is dense in X if and only if

(a) A is a Nachbin polynomial algebra, and

(b) for every finite rank continuous linear mapping $\Pi: E \to E$, with $\|\Pi\| \leq C$, and every $g \in A$, the $g \circ \Pi$ belongs to the t_b^m-closure of A.

Proof

If A is dense it is readily verified that (a) and (b) hold. Now to prove the converse, let A be a polynomial algebra satisfying (a) and (b). From the lemmas 1 and 2 in Prolla [51] it follows that the set $M = \{\phi(f) \mid \phi \in F^*, f \in A\}$ is a subalgebra of $C^m(E)$ such that $M \otimes F \subset A$. Further it is verified that if E_0 is any finite dimensional subspace of E then $M|E_0 \subset C^m(E_0)$ is a Nachbin algebra. Thus applying the Nachbin's theorem noted in definition 4.3.12 it follows that $M|E_0$ is dense in $C^m(E_0)$ w.r.t. uniform convergence of a function and its first m derivatives on compact sets in E_0. It follows from this that $M|E_0 \otimes F$ is dense in $C^m(E_0,F)$.

To complete the proof let $f \in X$. Let $\epsilon > 0$, and B be a bounded set in E, and $j \in \mathbb{N}$, $j \leq m$. By the preceding lemma it follows that there is a finite rank continuous linear mapping $\Pi \in E^* \otimes E$ with $\|\Pi\| \leq C$ such that

(1) $\|D^i f(x) - D^i(f \circ \Pi)(x)\| < \epsilon/3$ if $x \in B$ and $i \leq j$. Since (B) is relatively compact subset of E, by the observations made in the preceding paragraph it is verified that there is $g \in A$ such that

(2) $\|D^i(f|E_0)(t)(y) - D^i(g|E_0)(t)(y)\| < \delta$ $(t \in \Pi(B)$, $y \in E_0 = \Pi(E)$, $\|y\| \leq 1$, $i \leq j)$, where $\delta > 0$ is sufficiently small depending on ϵ, and C, and j. By the chain rule it follows that

(3) $\|D^i f(\Pi(x)) \circ \Pi - D^i g(\Pi(x)) \circ \Pi\| < \epsilon/3$ if $x \in B$ and $i \leq j$. Finally using the condition (b) of the theorem it follows that there is a $g \in A$ such that

(4) $\|D^i(g \circ \Pi)(x) - D^i h(x)\| < \epsilon/3$, $x \in B$, $i \leq j$.

From (1), (3), and (4) we have

$$\|D^i f(x) - D^i h(x)\| < \epsilon \quad \text{if} \quad x \in B, \quad i \leq j,$$

completing the proof.

We have the following analogue of Bernstein's theorem for infinite dimensional spaces.

4.3.17 Corollary

Let E,F be as in the preceding theorem. Then $P_f(E,F)$ is dense in $C^m_W(E,F)$ for all $m \geq 1$.

Proof

This is deduced noting that $P_f(E,F)$ is a Nachbin polynomial algebra and $P \circ \Pi \in P_f(E,F)$ for all $P \in P_f(E,F)$ and $\Pi \in E^* \otimes E$.

It is noted here that a special case of the preceding corollary has been proved in [54] for reflexive E under a certain condition which is stronger than the assumption made here on E i.e. E^* has B.A.P.

4.4 Analytic approximations in Banach spaces

A well known theorem of Whitney [65] in part implies that if G is an open subset of real n-dimensional Banach space E_n, and $\phi(x)$ is a positive continuous function in G, and if $f(x)$ is a continuous real valued function, then there is an analytic function $h(x)$ on E_n such that

$$|f(x) - h(x)| \leq \phi(x).$$

Kurzweil [37] discussed the generalization of this approximation theorem to arbitrary real separable Banach spaces. From the results in 3.2, it follows that any Banach space which has the preceding approximation property is superreflexive. The results of Kurzweil while implying that not all superreflexive spaces have this property, provide a very general sufficient condition on the geometric structure of the Banach space, assuring the approximation property.

Before proceeding to Kurzweil's theorem we recall a few basic definitions. All polynomials on Banach spaces considered in this section are continuous.

4.4.1 Definition

Let G be an open subset of real Banach space E and B be a Banach space. A function $f: G \to B$ is said to be analytic if for each $x \in G$, there exists an open ball $B_x \subset G$, and homogeneous polynomials P_n^x of degree n on E into B, such that

$$f(x+h) = \sum_{n \geq 0} P_n^x(h^{(n)}), \quad \text{if} \quad x+h \in B_x$$

where the series on the right side converges in the norm topology of E to $f(x+h)$.

For a discussion of analytic functions on Banach spaces, see Hille and Phillips [26].

As usual we associate with each real Banach space its complexification $\tilde{E}$, where $\tilde{E}$ is defined as follows:

$$\tilde{E} = \{x + iy \mid (x,y) \in E \times E\}.$$

The vector space $\tilde{E}$ is equipped with the norm $|||x + iy||| = \sup_{0 \leq \alpha \leq 2\pi} \|x \cos \alpha - y \sin \alpha\|$ where $\| \ \|$ is the norm of the space E.

We state without proof a known extension lemma for polynomials, stated in [2].

4.4.2 Lemma

If $p(x)$ is a real polynomial on E, then there is a polynomial $\tilde{p}(x)$ defined on $\tilde{E}$ such that $\tilde{p}(z) = p(z)$ if $z = x + iy$ with $y = 0$.

4.4.3 Definition

A Banach space E is said to have the polynomial support property if there is a polynomial $p(x)$ of degree ≥ 1, such that

$$p(0) = 0, \quad \text{and} \quad \inf\{|p(x)| \mid \|x\| = 1\} > 0.$$

4.4.4 Remark

It is clear that the polynomial $p(x)$ in the preceding definition is at least of degree 2. Further if $p(x) = p_1(x) + P_2(x) + \ldots + p_m(x)$, where $p_i(x)$ are homogeneous polynomials of degree i, satisfying the inequality in the definition, then for the polynomial $q(x) = \sum_{i=1}^{m} p_i^2(x)$, $q(0) = 0$, and $\inf_{\|x\|=1} q(x) > 0$. Thus we may assume in the preceding definition that $p(x)$ is non-negative.

4.4.5 Lemma

If $p(x)$ is a non-negative polynomial on a Banach space E such that $\inf_{\|x\|=1} p(x) > 0$, then

(i) if $K(y,r) = \{x | p(x-y) < r\}$, $r > 0$, then $K(y,r)$ is a bounded open neighborhood of x,

(ii) the sets $\{K(y,r) | r > 0\}$ is a fundamental system of neighborhoods for the norm topology of E, at x.

The lemma follows from the definition of a polynomial and the hypotheses on $p(x)$.

4.4.6 Lemma

If E is a separable Banach space, and if $p(x)$ is as in the preceding lemma, and if $\{K(x,r(x))\}$ is a covering of an open set $G \subset E$, then there exists locally finite open coverings $\{D_i\}_{i \geq 1}$, $\{D_i^*\}_{i \geq 1}$ of G such that

$$D_i \subset K(x_i, r(x_i)), \; D_i^* \subset K(x_i, r(x_i) + 2\epsilon_i)$$

where $\{x_i\}$ is a sequence in G, and $\{\epsilon_i\}_{i \geq 1}$, $\{r(x_i)\}_{i \geq 1}$ are sequences of positive numbers with

$$\epsilon_i \to 0, \text{ and } 3\epsilon_i < r(x_i).$$

Proof

Since E is separable we can choose a countable covering $\{K(x_i, r(x_i)\}$ of G with $x_i \in G$. Choose a sequence of positive numbers $\{\epsilon_i\}$, $\epsilon_i \to 0$, $1 > \epsilon_1 > \epsilon_2 \ldots$ and $3\epsilon_i < r(x_i)$ for all $i \geq 1$. Let the sequence of sets $\{D_i\}$ be defined as follows. Let

$$D_1 = K(x_1, r(x_1)).$$

Let

$$D_n = [\prod_{i=1}^{n-1} C(x_i, r(x_i) - 3\epsilon_i)] \cap K(x_n, r(x_n)), \quad n \geq 2,$$

where

$$C(x,r) = \{y \mid p(y - x) > r\}.$$

If $y \in G$, choose the smallest integer m, such that

$$y \in K(x_m, r(x_m)), \quad y \notin \bigcup_{i=1}^{m-1} K(x_i, r(x_i)).$$

Further there is an index $\ell > m$ such that $y \in K(x_m, r(x_m) - 3\epsilon_\ell)$ since $\epsilon_\ell \to 0$. From the choice of $\{\epsilon_i\}$,

$$K(x_m, r(x_m) - 3\epsilon_\ell) \cap C(x_m, r(x_m) - \epsilon_j) = \emptyset$$

for all $j > \ell$. Thus there is a neighborhood $K(x_m, r(x_m) - 3\epsilon_\ell)$ that intersects only finitely many of the sets D_i. Hence $\{D_i\}$ is a locally finite open covering with D_i as desired. We proceed to define the sets $\{D_i^*\}_{i \geq 1}$.

Define the sets

$$D_n^* = [\prod_{i=1}^{n-1} C(x_i, r(x_i) - 3\epsilon_n)] \cap K(x_n, r(x_n) + 2\epsilon_n) \quad \text{for } n = 1,2,3,\ldots.$$

It is verified that $D_n \subset D_n^* \subset G$, for all $n \geq 1$, and $\{D_n^*\}$ is a covering of G. The sets $K(x_n, r(x_n) - 3\epsilon_\ell)$, $\ell > n$, are in D_j^*, $j \geq \ell$. Hence $\{D_j^*\}$ is a locally finite open covering of G. Clearly the coverings $\{D_j^*\}$, $\{D_j\}$ satisfy all the conditions in the lemma.

4.4.7 <u>Theorem</u>

If E is a separable real Banach space with the polynomial support property, and G is an open subset of R, then if F is a continuous mapping on $G \to E_1$, where E_1 is another Banach space, there exists an analytic mapping H on $G \to E_1$ such that

(A) $$\|F(x) - H(x)\| < 1, \quad \text{for all } x \in G.$$

Proof

Let $p(x)$ be a non-negative polynomial on E such that $\inf_{\|x\|=1} p(x) > 0$. Let $K(x,r)$, $C(x,r)$ be as in Lemma 4.4.5. For each $x \in G$, choose a positive number $r(x)$ such that $K(x,2r(x)) \subset G$, and for all $y \in K(x,2r(x))$, $\|F(x) - F(y)\| < \frac{1}{4}$. The sets $\{K(x,r(x)\}$ is an open covering of G, and by the preceding lemma we can construct two locally finite coverings $\{D_j\}$, $\{D_j^*\}$ with properties stated in the lemma.

Let E_n be the n-dimensional Euclidean space. Let the sets $T_n \subset E_n$ be defined for $n \geq 1$ as follows:

$$T_1 = \{e_1 | -1 \leq e_1 \leq r(x_1) + \epsilon_1\}$$

and

$$T_n = \{(e_1, e_2, \ldots, e_n) | r(x_i) - 2\epsilon_n \leq e_i \leq V_n,\ i = 1,2,\ldots,n-1 \text{ and } -1 \leq e_n \leq r(x_n) + \epsilon_n\}, \text{ for all } n \geq 2, \text{ where}$$

$$V_n = \sup\{p(x-x_i) + 2 | x \in K(x_n, 2r(x_n)),\ 1 \leq i \leq n-1\}, \quad n \geq 2.$$

Let functions ϕ_n be defined on $\tilde{E}$, by setting

$$\phi_n(z) = C_n(1 + \|F(x_n)\|)\int_{T_n} \exp[-t_n \Sigma a_i(\tilde{p}(z-x_i) - e_i)]d\mu_n$$

where μ_n is the n-dimensional Lebesgue measure on E_n, and

$$\frac{1}{C_n} = \int_{E_n} \exp[-t_n(\sum_{i=1}^{n} a_i e_i^2)]d\mu_n = b_n t_n^{-n/2}, \quad n \geq 1.$$

Here b_n, $n \geq 1$ are positive constants depending on $a_1, a_2, \ldots$ and $t_1, t_2, \ldots$ which will be chosen to satisfy certain constraints. Since the functions ϕ_j are composites of the analytic function $A: \tilde{E} \to C^j$, where $A(z) = (\tilde{p}(z-x_1), \tilde{p}(z-x_2), \ldots, \tilde{p}(z-x_j))$, and C^j is the j-dimensional complex space, and of an analytic function on C^j, ϕ_j are analytic on $\tilde{E}$ for all $j \geq 1$. Choose the positive numbers a_n so that if $\deg p = m$,

(1) $\sum_{n\geq 1} a_n(1+\|x-x_n\|)^{4m}$ converges for all $x \in E$.

For example, choose $a_n = \dfrac{1}{n!(1+\|x_n\|)^{4m}}$. Further from the definitions of D_n, D_n^*, and T_n, $n \geq 1$, we can choose t_n so that

(2) $$t_n \geq (n!)^2(1+\|F(x_n)\|)^2\mu_n(T_n)+1,$$

(3) $$\left|\phi_n(x) - \|F(x_n)\| - 1\right| < 1/2 \quad \text{if} \quad x \in D_n,$$

and

(4) $$|\phi_n(x)| < \frac{1}{2^{n+3}(1+\|F(x_n)\|)} \quad \text{if} \quad x \notin D_n^*.$$

Now let $\phi(x) = \sum_{i\geq 1} \phi_i(x)$, and $H^*(x) = \sum_{i\geq 1} F(x_i)\phi_i(x)$.

Since uniform limit of analytic operations in complex Banach spaces are analytic, it follows that ϕ, and H^* are analytic functions on $\tilde{E}$. Assuming (*) that for every $x_0 \in G$ there is a positive number δ, and a positive integer n_0 such that

$$(1+\|F(x_n)\|)\phi_n(x_0+z) < \frac{1}{2^n}$$

for $z \in \tilde{E}$, $\|z\| < \delta$, $n > n_0$, let us complete the proof of the theorem.

Let $H(x) = \dfrac{H^*(x)}{\phi(x)}$. The function H is analytic on $\tilde{E}$, and we prove that it satisfies the requirement (A) in the theorem. Let $x \in G$. Then

$$F(x) - H(x) = F(x)\sum_{i\geq 1}\frac{\phi_i(x)}{\phi(x)} - \sum_{i\geq 1}\frac{F(x_i)\phi_i(x)}{\phi(x)}.$$

Let I_1 (I_2) denote the set of indices j such that $x \in D_j^*$ and ($x \notin D_j^*$).

$$\|F(x)-H(x)\| \leq \frac{1}{\phi(x)}\sum_{j\in I_1}\|F(x)-F(x_j)\|\,\phi_j(x)$$
$$+\frac{\|F(x)\|}{\phi(x)}\sum_{j\in I_2}\phi_j(x) + \frac{1}{\phi(x)}\sum_{j\in I_2}\|F(x_j)\|\,\phi_j(x).$$

If $j \in I_1$, then $x \in D_j^* \subset K(x_j, 2r(x_j))$, so that $\|F(x)-F(x_j)\| < \frac{1}{4}$.
Since $x \in D_\ell$ for some ℓ,

$$\|F(x) - F(x_\ell)\| < \frac{1}{4}, \quad \phi_\ell(x) > \|F(x_\ell)\| + \frac{1}{2}$$

and

$$\phi(x) \geq \phi_\ell(x) > F(x), \quad \phi(x) > \frac{1}{2}, \quad \frac{\|F(x)\|}{\phi(x)} < 1.$$

Now since

$$\sum_{j \in I_2} \phi_j(x) \leq \sum_{j \geq 1} \frac{1}{2^{j+3}} = \frac{1}{8},$$

and

$$\sum_{j \in I_2} \|F(x_j)\| \phi_j(x) \leq \frac{1}{8},$$

it follows that $\|F(x) - H(x)\| \leq \frac{1}{4} + \frac{1}{8} + \frac{2}{8} < 1$, completing the proof of the theorem. All that remains is the verification of (*).

Fix $x_0 \in G$. There is an index j_0 such that $x_0 \in K(x_{j_0}, r(x_{j_0}))$, $x_0 \notin K(x_j, r(x_j))$, $j = 1,2,\dots,j_0-1$. Thus there is a positive number α, and a positive integer n^1 such that $e_{j_0} - p(x_0 - x_{j_0}) > \alpha$ for every vector $(e_1,e_2,\dots,e_n) \in T_n$, $n > n^1$. Since p is a polynomial of degree $2m$, we have

$$\tilde{p}(x_0 - x_j + z) = p(x_0 - x_j) + \xi_j$$

where $|\xi_j| \leq M[1 + \|x - x_j\|]^{2m} \|z\|$, if $\|z\| \leq 1$, see Whitney [65], for some constant M. Thus, since

$$[\tilde{p}(x_0 + z - x_j) - e_j]^2 = [p(x_0 - x_j) - e_j + \xi_j]^2,$$

it follows that

$$\mathrm{Re}[\tilde{p}(x_0 + z - x_j) - e_j]^2 \geq [|p(x_0 - x_j) - e_j| - |\xi_j|]^2 - 2|\xi_j|^2.$$

Thus

$$\begin{aligned}
&\mathrm{Re}\{\sum_{j=1}^{n} a_j(p(x_0 + z - x_j) - e_j)^2\} \\
&\geq -2\ \Sigma a_j|\xi_j|^2 + a_{j_0}(|p(x_0 - x_j) - e_{j_0}| - |\xi_{j_0}|)^2 \\
&\geq -2M^2\|z\|^2 \sum_{j\geq 1} a_j(1+\|x_0 - x_j\|)^{4m} + \\
&\quad + a_{j_0}\{\alpha - M(1 + \|x_0 - x_j\|)^{2m}\|z\|\}^2,
\end{aligned}$$

where $(e_1,e_2,\dots,e_n) \in T_n$ and α is positive. As the series

$\sum_{j\geq 1} a_j(1 + \|x_0 - x_j\|)^{4m}$ is convergent, it is possible to choose β and δ such that the inequality

$$\mathrm{Re}\{\sum_{j=1}^{n} a_j(p(x_0 + z - x_j) - e_j)^2\} > \beta$$

holds for $z \in \tilde{E}$, $\|z\| < \delta$, $(e_1,\ldots,e_n)$ T_n, $n > n^1$. From this it is verified that

$$(\|F(x_n)\| + 1)|\phi_n(x_0+z)| \leq \frac{t_n^{n/2}}{b_n}(\|F(x_n)\| + 1)^2 \mu_n(T_n)e^{-\beta t_n}$$

$$\leq \frac{(\|F(x_n)\| + 1)^2 \mu_n(T_n)}{b_n} \cdot \frac{n!}{\beta^n t_n^{n/2}}$$

$$\leq \frac{1}{n!\ \beta^n},$$

so that there is an integer n_0 such that for $n > n_0$, $\|z\| < \delta$, the inequality (*) holds, completing the proof of the theorem.

We deduce from the preceding theorem the following theorem.

4.4.8 Theorem

If E is a Banach space as in the preceding theorem, and F is a continuous mapping on an open set G in E into a Banach space E_1, and ϕ is a continuous positive function on G, then there is an analytic function $H:G \to E_1$ such that

$$\|F(x) - H(x)\| < \phi(x).$$

Proof

By the preceding theorem there is an analytic map $\psi:G \to E_1$ such that

$$\left|\frac{1}{\phi(x)} + 1 - \psi(x)\right| < 1 \quad \text{for all} \quad x \in G.$$

Clearly $\psi(x) > \frac{1}{\phi(x)}$. By the same token there is an analytic map $H^*G \to E_1$ such that

$$\|\psi(x)F(x) - H^*(x)\| < 1, \quad \text{for all} \quad x \in G.$$

Let $H(x) = \frac{H^*(x)}{\phi(x)}$. Then H is analytic on $G \to E_1$ and satisfy the requirements, completing the proof of the theorem.

As a corollary we obtain the following result.

4.4.9 Theorem

Let E be a Banach space such that $\ell_p(L_p[0,1])$ is isomorphic with a subspace of E, Then if p is not an even integer, E does not enjoy the analytic approximation property.

Proof

Let $g: R^+ \to R$ be a continuous function such that $g(\xi) = 1$ if $0 \leq \xi \leq 1$, $g(\xi) = 0$ if $\xi \geq 2$, $0 \leq g \leq 1$. Define $F: E \to R$ by setting $F(x) = g(\|x\|)$. Then F is a continuous function on E. Let $\phi: E \to R$ be a non-negative continuous function with $\phi(0) = 1$, and $\phi(x) = 0$ if $\|x\| \geq 1$. Since E has the analytic approximation property there is an analytic function $H: E \to R$ such that

$$|F(x) - H(x)| \leq \phi(x), \quad \text{for all } x \in E.$$

From the choice of F, and ϕ it is verified that the support of H is bounded. Thus H is a C^∞-function with bounded support on E. From the theorem 3.3.3 it follows that p is an even integer. This contradiction completes the proof.

4.5 Simultaneous approximation of smooth mappings

The purpose of this section is to briefly recall certain recent results concerning the simultaneous approximation of a C^k-mapping from a Banach space E into another Banach space F by functions of class C^p where $p > k$. Before stating the results precisely we state a definition.

4.5.1 Definition

Let E, F be Banach spaces, and G be an open subset of E. As usual let us denote the vector space of C^k-mappings on G into F by $C^k(G,F)$. Then the sets

$$N(f, \epsilon(\cdot)) = \{g \in C^k(G,F) \mid \|D^i g(x) - D^i f(x)\| < \epsilon(x),\ 1 \le i \le k \text{ for all } x \in G\}$$

where $f \in C^k(G,F)$, and $\epsilon(x)$ is an arbitrary continuous function on $E \to R$ constitute a basis for a topology on $C^k(G,F)$. This definition extends in a natural way to $C^k(M,N)$ where M,N are C^∞-manifolds modelled on the Banach spaces E and F respectively.

Inspired by certain simultaneous theorems of Eells and McAlpin, Smale and Quinn,(see Moulis [43]) Moulis proved the following theorem in [43].

Let E^α be any one of the Banach spaces ℓ_p or c_0, where p is an integer $2 \le p < \infty$ admitting an equivalent norm of class C^α away from 0.

4.5.2 Theorem

Let G be an open subset of E^α, and F be an arbitrary Banach space. The set of mappings of class C^α on G into F is dense in the space $C^1(G,F)$ equipped with C^1-fine topology.

The theorem has been further extended to manifolds modelled on Hilbert spaces, (see Appendix in this lecture notes), as follows.

4.5.3 Theorem

Let M, N be separable paracompact C^∞-manifolds modelled on the Hilbert spaces E and F. Let N_1 be a submanifold of N. Then the set of C^∞-mappings on M into N transversal to N_1 is dense in $C^1(M,N)$ endowed with C^1-fine topology.

Further generalizations of the theorem 4.5.2 has been considered by Heble in [24], and the following theorem has been proved in [24].

4.5.4 Theorem

Let G be an open subset of a Hilbert space H, and F be an arbitrary Banach space. Then the set $C^{\infty}(G,F)$ is dense in the space $C^k(G,F)$ in the C^k-fine topology, where $k \geq 0$.

For the proofs of the theorems 4.5.2 and 4.5.3, see [43] and for the proof of the theorem 4.5.4 see [24].

APPENDIX

Infinite Dimensional Differentiable Manifolds

A.0 Introduction

The theory of infinite dimensional differentiable manifolds has been studied by several mathematicians in different contexts. For an account of the theory, see Abraham and Robbin [1] , Kahn [30] , Lang [38] and Palais [49] . In this appendix we are concerned with a few results on the geometry of differentiable manifolds of considerable importance in nonlinear analysis which have appeared in journals and have not been dealt with in the monographs mentioned above. We include a generalization of Morse Lemma [47] and the theorems of Bessaga [7] and Eells-Elworthy [21] on diffeomorphisms. Before proceeding to these important results we recall some basic definitions concerning differentiable manifolds modelled on Banach spaces.

A.1 Preliminaries

A.1.1. Definition

Let E, F be Banach spaces, U and V be open subsets of E and F respectively and f a homeomorphism of class C^p, $0 \leq p \leq \infty$, of U onto V. If the inverse of f is also of class C^p, then f is said to be a C^p-isomorphism.

The following characterization of C^p-isomorphisms in terms of C^1-isomorphisms is a consequence of the fixed point theorem on contractions.

A.1.2. Theorem

Let E, F, U and V be as in Definition A.1.1. If f is a C^p-homeomorphism on U onto V such that f is C^1-isomorphism, then f is a C^p-isomorphism.

Proof

We prove first that f is a local C^p-isomorphism at each point $x_0 \in U$.

Since a linear isomorphism is a C^∞-isomorphism, we may assume without loss of generality that $E = F$ and the derivative $f'(x_0)$ is the identity, by replacing if necessary f by $(f'(x_0))^{-1}f$. Further since translations are C^∞-isomorphisms, we may assume $x_0 = 0$, and $f(x_0) = 0$. Let $g(x) = x - f(x)$. Then $g'(x_0) = 0$, and since g' is continuous there is a positive number a such that if $\|x\| < 2a$, then $\|g'(x)\| < \frac{1}{2}$. From the mean value theorem it follows that $\|g(x)\| \leq \frac{1}{2}\|x\|$, and g maps the closed ball of radius a, $\overline{B_a(0)}$ into $\overline{B_{a/2}(0)}$. Let now $y \in \overline{B_{a/2}}(0)$. We claim that there is a unique element $x \in \overline{B_a}(0)$ such that $f(x) = y$. To this end consider the map $g_y(x) = y + x - f(x)$. If $\|y\| \leq a/2$ and $\|x\| \leq a$, then $\|g_y(x)\| \leq a$. Hence g_y is a mapping on the complete metric space $\overline{B_a}(0)$ into itself. Since $\|g'(x)\| < 1/2$ if $\|x\| < 2a$, it follows from the mean value theorem that

$$\|g_y(x_1) - g_y(x_2)\| = \|g(x_1) - g(x_2)\| \leq \frac{1}{2}\|x_1 - x_2\|,$$

for x_1, x_2 in $\overline{B_a(0)}$. Thus g_y is a contraction. Hence g_y has a unique fixed point. Thus there exists only one element x such that $f(x) = y$. Let the local inverse f^{-1} be denoted by ϕ. Further

$$\begin{aligned}\|x_1 - x_2\| &\leq \|f(x_1) - f(x_2)\| + \|g(x_1) - g_2(x)\| \\ &\leq \|f(x_1) - f(x_2)\| + \tfrac{1}{2}\|x_1 - x_2\|.\end{aligned}$$

Hence $\|x_1 - x_2\| \leq 2\|f(x_1) - f(x_2)\|$, and

it follows that ϕ is continuous. We verify that ϕ is differentiable in $B_{a/2}(0)$. Let $y_i = f(x_i)$, $i = 1,2$, with $y_i \in B_{a/2}(0)$ and $x_i \in \overline{B_a(0)}$. Thus

$$(*) \qquad \|\phi(y_1) - \phi(y_2) - (f'(x_2))^{-1}(y_1 - y_2)\|$$

$$= \|x_1 - x_2 - (f'(x_2))^{-1}(f(x_1) - f(x_2))\|$$

Using the identity operator in the form $(f'(x_2))^{-1}f'(x_2)$ inside the norm on the right hand side of (*) and applying the Schwarz inequality, each norm is bounded by $\|f'(x_2)^{-1}\|\|f'(x_2)(x_1-x_2) - f(x_1) + f(x_2)\|$. From the differentiability of f, noting that ϕ is continuous it is verified that ϕ is differentiable, and $\phi'(y) = f'(\phi(y))^{-1}$, for $y \in B_{a/2}(0)$. Since f', and the inverse operation are continuous it follows that ϕ' is continuous. Thus ϕ is of class C^1. Since taking inverse operation and f' are of class C^{p-1}, it follows that ϕ is of class C^p, completing the proof.

A.2 Differentiability in a half-space

A.2.1 Definition

If E is a Banach space, $f \in E^*$, $f \neq 0$, then $E_f^+ = \{x | x \in E,\ f(x) \geq 0\}$ is called the (positive) half-space defined by f. $\partial E_f^+ = \{x | x \in E,\ f(x) = 0\}$ is called the boundary of the half-space E_f^+.

A.2.2 Definition

If H is a half-space in a Banach space E and U an open subset of H, $p \in U \cap \partial H_0$ then a function $f: U \to F$, (where F is another Banach space), is said to be a C^k-map if there is a neighborhood V of p in E and a C^k-map $g: V \to F$ such that $f|U \cap V = g|U \cap V$. The successive differentials $D^m g$ are well-defined at p and we define $D^m f(p) = D^m g(p)$. The map f is said to be C^k in U if it is C^k at each point of $U \sim \partial H$ in the usual sense and it is C^k at each point in $U \cap \partial H$ in the above sense.

If H_1 is a half-space of F and U' is an open set in H_1, then a bijection $f: U \overset{onto}{\to} U'$ is a C^k-isomorphism if f, f^{-1} are C^k-maps.

A.2.3 Theorem (Invariance of boundaries)

Let H_i, $i = 1,2$ be half-spaces in the Banach spaces E and F

respectively, and U_i , $i = 1,2$ be open subsets of H_i. Let $f_i : U_1 \longrightarrow U_2$ be a C^k-isomorphism, $k > 1$. Then f maps $U_1 \cap \partial H_1$ onto $U_2 \cap \partial H_2$.

The theorem is a consequence of the inverse function theorem.

A.3 Differentiable manifolds modelled on Banach spaces

We proceed now to the definition of differentiable manifolds modelled on Banach spaces.

A.3.1. Charts and Atlases

Let X be a Hausdorff topological space. A chart in X is a homeomorphism ϕ on an open set U of X onto an open subset of a Banach space E , or onto an open set in a half-space of E. Sometimes, the pair (Domain ϕ, ϕ) is referred to as a chart.

If ϕ , ψ are charts on X and U = Domain $\phi \cap$ Domain ψ , then ϕ , ψ are said to be C^k-compatible (or C^k-related) if $\psi \circ \phi^{-1}$ is a C^k-isomorphism of $\phi(U)$ onto $\psi(U)$.

A C^k-atlas for X is a collection $\mathcal{U}$ of charts in X which are pairwise C^k-compatible such that $X = \bigcup_{\phi \in \mathcal{U}}$ Domain ϕ . A complete C^k-atlas is one which is maximal in the ordering by inclusion.

A.3.2 Remark

Let $\mathcal{U}$ be an atlas for X and let ϕ, ψ be charts in X each of which is C^k-related to each chart in $\mathcal{U}$. Then ϕ and ψ are C^k-related. If $\mathcal{U}$ is any atlas, then $\mathcal{U}$ is included in a unique complete C^k-atlas $\mathcal{U}^*$, where $\mathcal{U}^* = \{ \phi \mid \phi$ is a chart in X which is C^k-related to each chart in $\mathcal{U}\}$.

A.3.3 Definition

A C^k-manifold modelled on a Banach space E is a pair (X, $\mathcal{U}$) where (i) X is a Hausdorff space, (ii) $\mathcal{U}$ is a complete atlas for X, and

(iii) the range of each chart in $\mathcal{U}$ is in the Banach space E.

If X is an open subset of a Banach space E, then i, the identity map on X is a chart, and $\{(X, i)\}$ defines a unique C^∞-structure on X modelled on E. The space X with this C^∞-structure is called a local manifold.

A.3.4 Definition

If X, Y are C^p-manifolds, $0 \leq p \leq \infty$, modelled on Banach spaces E and F respectively, then a mapping $f : X \to Y$ is said to be a local C^p-isomorphism if for each x in X, there is a chart (U, ϕ) at x, (V, ψ) at f(x), with $f(U) \subset V$, such that the map (local representative of f) $\psi \circ f \circ \phi^{-1} : \phi(U) \to \psi(V)$ is a C^p-isomorphism on $\phi(U)$.

A.3.5 Definition

If X is a C^k-manifold modelled on the Banach space E, the boundary ∂X of X is defined by

$$\partial X = \{x \mid x \in X \text{ such that there is a chart } \phi \text{ at } x \text{ mapping Dom}(\phi) \text{ onto an open subset of a half-space H of E with } \phi(x) \in \partial H\}.$$

A.3.6 Remark

If X is a C^k-manifold modelled on the Banach space E and $x \in \partial X$, then if ϕ is any chart at x, ϕ maps Dom(ϕ) onto an open subset of a half-space H of E, and $\phi(X) \in \partial H$. Further, if (X, A) is a C^k-manifold where A is an atlas defining the manifold structure on X, then $\{\psi \mid \psi = \phi|\partial X, \phi \in A\}$ is a C^p-atlas on ∂X.

We remark further that if H is a half-space of the Banach space E, then $\{i_H\}$ is a C^k-atlas on H for the topological space H modelled on E, where i_H is the identity map on H. Thus H is a C^k-manifold modelled on E.

We proceed to the concept of a submanifold of a manifold.

A.3.7 Definition

Let X be a C^k-manifold modelled on E and Y be a subspace of X.

Then a chart ϕ for X is said to admit a restriction to Y if there is a closed linear subspace F of E such that $\phi|Y$ is a chart with range in F i.e. $\phi|Y$ should map homeomorphically $(\mathrm{Dom}\ \phi) \cap Y$ onto an open subset of F or a half-space of F.

The following lemma is a consequence of the definitions stated above.

A.3.8 Lemma

If X is a C^k-manifold and Y is a subspace of X and ϕ, ψ are charts of X admitting restrictions to Y then $\phi|Y$, $\psi|Y$ are C^k-compatible.

A.3.9 Definition

If Y is a subspace of a C^p-manifold X, then Y is called a C^p-submanifold of X if for each $y \in Y$ there is a chart of X at y admitting a restriction to Y.

A.3.10 Theorem

If Y is a C^k-submanifold of the C^k-manifold X, and

$$\mathcal{A} = \{\text{the charts of } X \text{ admitting restrictions to } Y\}$$

then $$\mathcal{A}_1 = \{\phi|Y \,\big|\, \phi \in \mathcal{A}\}$$

is a C^k-atlas for Y. The manifold structure obtained by completing the atlas $\mathcal{A}_1$, (see A.3.1) defines the C^k-submanifold structure on Y.

In passing we note that if Y is a C^k-submanifold of the C^k-manifold X, then the inclusion $i_Y : Y \to X$ is a C^k-isomorphism. If X, X' are two C^k-manifolds and $f: X \to X'$ is a C^k-isomorphism, then $f|Y = f \circ i_Y$ is a C^k--isomorphism.

A.3.11 Remark

If X is a C^k-manifold and Y is an open subset of X, then Y is a C^k-submanifold. Further if X is a manifold with boundary then ∂X is a C^k-submanifold.

A.3.12 Definition

Let X_1, X_2 be two manifolds and $f : X_1 \to X_2$ be a morphism. f is said to be an immersion at $x_1 \in X_1$ if there is a neighborhood U_1 of x_1 such that $f|U_1$ is an isomorphism onto a submanifold of X_2; it is called an immersion if it is an immersion at all points of X_1. An immersion which is an isomorphism onto a submanifold is an embedding.

A morphism $f : X_1 \to X_2$ is a submersion at $x_1 \in X_1$ if there are charts (U, ϕ) at x_1, (V, ψ) at $f(x_1)$ such that ϕ is an isomorphism on U onto $U_1 \times U_2$, where U_1 and U_2 are open sets in some Banach spaces and such that the map $f_{V,U} = \psi \circ f \circ \phi^{-1} : U_1 \times U_2 \to V$ is a projection. If f is a submersion at all points of X_1, then it is called a submersion of X_1 into X_2.

A.3.13 Proposition

Let X_1, X_2 be two C^k-manifolds modelled on Banach spaces. Let $f: X_1 \to X_2$ be a C^k-morphism. Then (1) f is an immersion at a point x in X_1 if and only if there is a chart (U,ϕ) at x, a chart (V,ψ) at f(x) such that the map $f'_{V,U}(\phi(x))$ is injective and splits. (2) f is a submersion at x if and only if there are charts as in (1) such that $f'_{V,U}(\phi(x))$ is surjective and its kernel splits. (See [38], ch.2, sec.2.)

The concepts of vector bundles and tangent bundles to a differentiable manifold, which are of considerable importance in finite dimensional differentiable manifolds, admit natural generalizations when the manifold is modelled on a Banach space. See [38].

A.3.14 Example

The unit sphere S of a real Hilbert space H is a closed submanifold of H.

Proof

Let $x \in S$ and L be the orthogonal complement of the manifold generated by {x}. Let U be the open subset of H containing all vectors y with

$(x,y) > 0$, and $\|y\|^2 > \frac{1}{4}$ and $\|P_L y\| < \frac{1}{2}$, where P_L is the orthogonal projection of H onto L. Since P_L is an open map, it is verified that U is open.

Let $\sigma: U \to L \times R$ be the map $\sigma(y) = (P_L y, \|y\|^2)$. If $(h, \lambda^2) \in L \times R$, with $\|h\| < \frac{1}{2}$, $\lambda^2 > \frac{1}{4}$, there is only one vector $y \in U$, namely $y = h + \sqrt{\lambda^2 - \|h\|^2}\, x$, such that $\sigma(y) = (h, \lambda^2)$. Thus σ is a bijection on U onto $V_1 \times V_2$, where $V_1 = \{h \mid h \in L, \|h\| < \frac{1}{2}\}$, and $V_2 = \{a \mid a > \frac{1}{4}\}$. Further since the map $y \to \|y\|^2$ is a C^∞-map on H into R, it follows that σ is a C^∞-isomorphism on U onto $V_1 \times V_2$ which is an open subset of $L \times R$. Finally

$$\sigma(U \cap S) = V_1 \times \{1\}.$$

Thus S is a submanifold of H. Since S is a closed subset of H, the proof is completed.

We conclude this section recalling the fundamental theorems of Bessaga [7] and Eells and Elworthy [21] on diffeomorphisms of infinite dimensional manifolds.

If H is an arbitrary Hilbert space, Klee [32] has shown that the unit sphere of H is homeomorphic with H. Subsequently the natural extension whether the homeomorphism can be replaced by diffeomorphism has been discussed in [7]. The answer is in the affirmative and the proof is the content of [7]. In what follows we consider the unit sphere S of H as a submanifold of H as described in the preceding example.

A.3.15 Theorem [Bessaga]

Let S and H be as in the preceding example. Then S is C^∞-diffeomorphic with H.

The proof of the theorem is accomplished in several steps.

A.3.16 Proposition

There exists a C^∞-isomorphism h from H onto $H \sim \{0\}$, such that

$h(x) = x$ if $\|x\| \geq \frac{1}{2}$.

We provide the proof of the proposition after noting a few lemmas.

A.3.17 Lemma

There exists an incomplete norm $N(\cdot)$ on H such that

(i) $N(x) \leq \|x\|$, with N of class C^∞ on $H \sim \{0\}$,

(ii) a point $\tilde{x}$ in the completion of (H,N) with $\tilde{x} \notin H$,

(iii) a function $p:(0,\infty) \to \{x | x \in H,\ N(x) \leq \frac{1}{2}\}$ of class C^∞, such that $p(t) = 0$ if $t \geq 1$, $\lim_{t \to 0} N(p(t) - \tilde{x}) = 0$, $N(p'(t)) < \frac{1}{2}$ for all $t > 0$.

Proof

Choose any countable normalized orthogonal system $\{e_n\}_{n \geq 1}$ in H. For each n let ϕ_n be a C^∞ nonincreasing, real valued function on R such that $\phi_n(t) = 1$ if $t \leq 1/2n$, and $\phi_n(t) = 0$ if $t > 1/2n-1$. Let

$$d_n = \mathrm{Max}(2^{n+1},\ 2 \sup \phi_n'(t)),$$

$$p(t) = \sum_{n \geq 1} \phi_n(t)\, e_n,$$

$$N(x) = \Big(\sum_{n \geq 1} \frac{(x,e_n)^2}{d_n^2} + \|x - \sum_{n \geq 1} (x,e_n)e_n\|^2 \Big)^{1/2}$$

and finally

$$\tilde{x} = \sum_{n \geq 1} e_n .$$

Note that the series $\sum_{n \geq 1} e_n$ is a Cauchy series with respect to $N(x)$. It is at once verified that $N(x) \leq \|x\|$ and all the conditions of the lemma are fulfilled.

A.3.18 Lemma

If p, and N are chosen as in the preceding lemma, then the mapping

$$(1) \qquad h_1(x) = p(N(x)) + x$$

is a C^∞-isomorphism on $H \sim \{0\}$ onto H with $h_1(x) = x$ for $N(x) \geq 1$.

Proof

Let y be any vector in H and $\phi(t) = N(y - p(t))$ if $t > 0$, and $\phi(0) = N(y - \tilde{x})$, where we denote the completion of the norm N also by N. It follows from the preceding lemma that

(2) $\phi: [0,\infty) \to [0,\infty)$ and

$$|\phi(t_1) - \phi(t_2)| \leq N(p(t_1) - p(t_2))$$
$$= N\int_{t_1}^{t_2} p'(t)dt \leq \int_{t_1}^{t_2} N(p'(t))dt$$
$$\leq |t_2 - t_1| \sup_{t_1 \leq t \leq t_2} N(p'(t)).$$

Hence

(3) $$|\phi(t_1) - \phi(t_2)| \leq \frac{1}{2}|t_1 - t_2|.$$

From (2) and (3) and Banach's contraction principle it follows that the equation $\phi(t) = t$ has a unique solution. Thus for each $y \in H$ there is a unique number $t(y)$ such that

$$N(y - p(t(y))) = t(y).$$

Further, since $y \in H$, $y \neq \tilde{x}$. Hence $t(y) \neq 0$. From this it is verified that h_1 is a 1-1 mapping. The properties of p and N imply that h_1 is of class C^∞. Let

$$\psi(y,t) = t - N(y - p(t(y)))$$

for any $y \in H$, $y - p(t(y)) \neq 0$, and the mapping ψ is differentiable in a neighborhood of $(y_0, t(y_0))$ in H $[0,\infty)$ for any vector y_0 in H. But (3) implies

$$\frac{\partial\psi(y,t)}{\partial t} \geq \frac{1}{2} > 0.$$

Hence as a consequence of the implicit function theorem, see 1.2 in [14], it follows that h_1 is a C^∞-isomorphism.

Before proceeding to the proof of the proposition A.3.16 we note the following fact. If $\lambda:(0,\infty) \to (0,\infty)$ be a nondecreasing C^∞-function $\lambda(t) = 0$ if $t \leq \frac{1}{2}$, $\lambda(t) = 1$ if $t \geq 1$, let for $x \in H$,

$$h_2(x) = [\lambda(\|x\|)\frac{\|x\|}{N(x)} + 1 - \lambda(\|x\|)]x$$

if $x \neq 0$, and $h_2(0) = 0$. Once again applying implicit function theorem as in the proof of the previous lemma it is verified that h_2 is a 1-1 mapping on H onto H and transformed the unit cell of H onto the unit cell of the normed space (H,N) and h_2 is of class C^∞.

Proof of Proposition A.3.16

If h_2 is a map as noted above and h_1 is the map defined in lemma A.3.18 it follows that the mapping

$$h(x) = \frac{1}{2}\, h_2^{-1}(h_1^{-1}(h_2(2x))$$

has all the required properties.

Let $x_0 \in S$ and $A = \{x_0\} \cup \{x \mid \|x - x_0\| > \frac{1}{2}\}$. Consider the atlas $\{S_i,\phi_i\}_{i\varepsilon A}$ on S, where, for each $i \varepsilon A$, $S_i = \{x|(x,i) > 0\}$, $\phi_i: S_i \to H_i = \{x|x \varepsilon H, (x,i) = 0\}$, the orthogonal projection of the half-sphere S_i onto the subspace H_i. This atlas is verified to be compatible with the C^∞-structure of S inherited from H; see example A.3.14.

A.3.19 Proposition

Let $x_0 \in S$, the unit sphere of the Hilbert space H. Then there is a C^∞-isomorphism on S onto $S \sim \{x_0\}$.

Proof

Define the map g on $S \to S$ by setting

$$g(x) = x \text{ if } (x,x_0) < \frac{\sqrt{3}}{2},$$

and

$$g(x) = \phi_{x_0}^{-1}(h(\phi_{x_0}(x)), \text{ otherwise.}$$

where h is as in the proposition A.3.16 and ϕ_{x_0} is as defined in the paragraph preceding the proposition. It is verified that g is a 1-1 mapping on S onto $S \sim \{x_0\}$, and if $i \neq x_0$, $i \in A$, then

$\phi_i \circ g \circ \phi_i^{-1}$ is the identity mapping on S and that the maps $\phi_{x_0}^{-1}\ g\ \phi_{x_0}$, $\phi_{x_0}\ g^{-1}\ \phi_{x_0}^{-1}$ are class C^∞. Hence g is a C^∞-isomorphism.

We now proceed to the proof of the theorem A.3.15.

Proof

Let P be the tangent hyperplane at $-x_0$ to the sphere S. Let π be the stereographic projection of $S \sim \{x_0\}$ onto P with vertex of projection at x_0. π is a C^∞-isomorphism on $S \sim \{x_0\}$ onto P. Let $f_1(x) = \pi(g(x) + x)$, where g is as in the proposition A.3.19. Then f_1 is a C^∞-isomorphism on S onto the subspace of H orthogonal to x_0. Since every subspace of H of codimension 1 is linearly homeomorphic with H, it follows that S is C^∞-diffeomorphic with H.

The methods used in proving the theorem can be applied to prove the following more general theorem.

A.3.20 Theorem

If X is any infinite dimensional normed space whose norm is of class C^k on $X \sim \{0\}$, then the unit sphere of X is C^k-diffeomorphic with a subspace of X of codimension 1.

After the appearance of Bessaga's theorem, Bourghelea and Kuiper [36] established that if the unit sphere S of the Hilbert space H is equipped with the real analytic structure it inherits from H, then there is an analytical isomorphism on S onto H, thus answering a problem raised in [7].

The theorem of Bessaga clearly illustrates that infinite dimensional manifolds differ very much from their finite dimensional counterparts. As a further example we mention a deep theorem of Eells and Elworthy [21] which in turn is a generalization of Bessaga's theorem. Before proceeding to the statement of the theorem we note that a differentiable manifold is said to be parallelizable if its tangent bundle is trivializable, see [38].

A.3.21 Theorem [Eells and Elworthy]

Let X be a separable metrizable C^∞-manifold modelled on a C^∞-smooth Banach space E of infinite dimension with a Schauder basis. If X is parallelizable then there is a C^∞-embedding of X onto an open subset of E.

Further generalizations of the preceding theorem are also stated in [21].

We conclude the appendix with a theorem of Palais [47,48].

The Morse lemma on the local behavior of a smooth function on a finite dimensional domain near a nondegenerate critical point has been found to be useful in solving variational problems. Palais [47, 48] has extended the theorem when the domain of the function is in an arbitrary Hilbert space and more generally when the domain is in an arbitrary Banach space.

A.3.22 Definition

Let f be a real valued function of class C^{p+2} defined on an open subset U of a Banach space E. If $x_0 \in U$, then x_0 is said to be a critical point if $Df(x_0) = 0$. If further the map $x \to D^2f(x_0)(x_i)$ is a linear isomorphism on $E \to E^*$ then x_0 is said to be a nondegenerate critical point of f.

A.3.23 Theorem [Morse-Palais]

Let f be a C^{p+2} function, $p \geq 1$, defined on a neighborhood 0 of a Hilbert space H. If $f(0) = 0$, and 0 is a nondegenerate critical point of f then there exists a neighborhood V of 0, a C^p-isomorphism on V into H, and an invertible symmetric operator A such that

$$f(x) = (A\phi(x),\ \phi(x)),$$

for $x \in V$.

For a proof and applications of the theorem, see [38].

References

1. R. Abraham, and J. Robbin, Transversal Mappings and Flows, Benjamin, New York, 1967.

2. A. Alexiewicz and W. Orlicz, Analytic operations in real Banach spaces, Studia Math. 14(1954), 57-78.

3. R. M. Aron, Approximation of differentiable functions on a Banach space, North Holland Math. Studies, 12(1977), 1-17.

4. R. M. Aron and J. B. Prolla, Polynomial approximations of differentiable functions on Banach spaces, J. für Reine. Angewandte. Math., 313(1980), 195-216.

5. M. S. Berger, Nonlinearity and Functional Analysis, Academic Press, New York, 1977.

6. C. Bessaga and A. Pełczynski, Selected Topics in Infinite Dimensional Topology, Polska Acad. Nauk., Warsaw, 1975.

7. C. Bessaga, Every Hilbert space is diffeomorphic with its sphere, Bull. L'Acad. Polonaise. Sci. 14(1966), 27-31.

8. E. Bishop and R. R. Phelps, A proof that every Banach space is subreflexive, Bull. Amer. Math. Soc. 67(1961), 97-98.

9. N. Bonic and J. Frampton, Smooth functions on Banach manifolds, J. Math. Mechanics 15(1966), 877-898.

10. S. H. Cox and S. B. Nadler, Supremum norm differentiability, Ann. Soc. Math. Polonaise 15(1971), 125-131.

11. M. M. Day, Normed Linear Spaces, Third Edition, Springer-Verlag, New York, 1973.

12. J. Diestel, Geometry of Banach Spaces - Selected Topics, Springer-Verlag Lecture Notes 485, 1975.

13. J. Diestel and J. J. Uhl, Vector Measures, Math. Surveys 15, Amer. Math. Soc., Providence, Rhode Island, 1977.

14. J. Dieudonné, Foundations of Modern Analysis, Revised Ed., Academic Press, New York, 1980.

15. J. Dieudonné, Treatise on Analysis, Vol. II, Academic Press, New York, 1970.

16. J. Dieudonné, Treatise on Analysis, Vol. III, Academic Press, New York, 1972.

17. N. Dinculeanu, Vector Measures, Pergamon Press, London, 1967.

18. S. Dinnen, Complex Analysis in Locally Convex Spaces, North Holland Math. Studies 59, 1981.

19. N. Dunford and J. T. Schwartz, Linear Operators, Part I, Interscience, New York, 1958.

20. J. Eells, A setting for global analysis, Bull. Amer. Math. Soc. 72(1966), 751-807.

21. J. Eells and K. Elworthy, Open embeddings of certain Banach manifolds, Annals of Math. 91)1970), 465-485.

22. P. Enflo, Banach spaces which can be given equivalent uniformly convex norms, Israel J. Math. 13(1972), 281-288.

23. P. Enflo, On the nonexistence of uniform homeomorphisms between L_p-spaces, Ark. Math. 8(1969), 103-105.

24. M. P. Heble, Approximation of differentiable functions on a Hilbert space, Mathematical reports of the Academy of Sciences of Canada 5(1983), 179-183.

25. M. P. Heble, Approximation of differentiable functions on a Hilbert space II, to appear.

26. E. Hille and R. S. Phillips, Functional Analysis and Semi-groups, Amer. Math. Soc. Colloquium Publications 31, Amer. Math. Soc., Providence, R.I., 1957.

27. R. C. James, Weak compactness and reflexivity, Israel J. Math. 2(1964), 101-119.

28. R. C. James, Some self-dual properties of normed linear spaces, Annals of Math. Studies 69(1972), 159-179.

29. M. I. Kadec, On linear dimension of L_p-spaces, Uspehi Math. Nauk., 13(1958), 95-98.

30. D. W. Kahn, Introduction to Global Analysis, Academic Press, New York, 1980.

31. T. Kato, Perturbation theory of nullity deficiency and other quanties of linear operators, J. Analyse. Math., 6(1958), 261-322.

32. V. Klee, Convex bodies and periodic homeomorphisms in Hilbert spaces, Trans. Amer. Math. Soc., 74(1953), 10-43.

33. V. Klee, Mappings into normed linear spaces, Fund. Math. 49 (1960), 25-63.

34. J. L. Krivine, Sous-espace de dimension finie des espaces de Banach réticulés, Ann. of Math. 104(1976), 1-29.

35. N. H. Kuiper, The homotopy type of the unitary group of a Hilbert space, Topology 3(1965), 19-30.

36. D.Bourghelea and N.H.Kuiper, Hilbert manifolds, Ann. of Math. 90(1969), 379-417.

37. J. Kurzweil, On approximation in real Banach spaces, Studia. Math. 14(1953), 214-231.

38. S. Lang, Introduction to Differentiable manifolds, Addison Wesley, Reading, Mass., 1972.

39. I. E. Leonard and K. Sundaresan, Geometry of Lebesgue-Bochner function spaces - smoothness, Trans. Amer. Math. Soc. 198(1974), 229-250.

40. J. Lesmes, On the approximation of continuously differentiable functions in Hilbert spaces, Rev. Coll. Math. 8(1974), 217-223.

41. J. G. Llavona, Approximation of continuously differentiable functions. Thesis, University Complutense, Madrid, 1975.

42. V. Z. Meshkhov, Smoothness properties in Banach spaces, Studia. Math. LXIII (1978), 111-123.

43. N. Moulis-Desolneux, Approximation de fonctions differentiables sur certain éspaces de Banach, Ann. Inst. Fourier 21(1971), 293-345.

44. L. Nachbin, Sur les algèbres denses de fonctions différentiables sur une variété , C. R. Acad. de Sci. 288(1949), 1549-1551.

45. E. Nelson, Topics in Dynamics I: Flows, Math. Notes, Princeton University Press, Princeton, N.J., 1969.

46. A. M. Nemirovskii and S. M. Semenov, On polynomial approximation in function spaces, Math. Sbornik. 21(1973), 255-277.

47. R. S. Palais, Morse theory on Hilbert manifolds, Topology 7 (1963), 299-340.

48. R. S. Palais, The Morse lemma for Banach spaces, Bull. Amer. Math. Soc. 75(1969), 968-971.

49. R. S. Palais, Foundations of Global Analysis, Benjamin, New York, 1968.

50. R. R. Phelps, A representation for bounded convex sets, Proc. Amer. Math. Soc. 11(1960), 976-983.

51. J. B. Prolla, On polynomial algebras of continuously differentiable functions, Atti.Accad.Naz. Lincei, 57(1974), 481-486.

52. J. B. Prolla and C. S. Guerreiro, An extension of Nachbin's theorem to differentiable functions on Banach spaces with the approximation property, Arkiv für Math. 14(1976), 251-258.

53. G. Restrepo, Differentiable norms in Banach spaces, Bull. Amer. Math. Soc. 70(1964), 413-414.

54. G. Restrepo, An infinite dimensional version of a theorem of Bernstein, Proc. Amer. Math. Soc. 23(1969), 193-198.

55. I. Singer, Best Approximation in Normed Linear Spaces, Springer-Verlag, New York, 1970.

56. J. Stern, Some applications of Model theory to Banach space theory, Annls. of Math. Logic 9(1976), 49-121

57. K. Sundaresan, Smooth Banach spaces Math. Ann. 173(1967), 191-199.

58. K. Sundaresan, Some geometric properties of the unit cell in spaces C(X,B), Bull. L'Acad. Polonaise. Sci. XIX (1971), 1007-1011.

59. K. Sundaresan, Geometry and nonlinear Analysis in Banach spaces, Pacific J. Math. 102(1982), 487-489.

60. J. Todd, Special polynomial in numerical analysis, Proc. Symposium on Num. Analysis, University of Wisconsin, Madison, Wisconsin (1968), 423-446.

61. N. Tomczak-Jaegermann, On the differentiability of the norm in the trace classes S_p, Seminaire Maurey-Schwartz, Ecole Polytech. Centre de Math. Exposé 22(1974-75).

62. H. Torunczyk, Smooth partitions of unity, Preprint 31, Polish Acad. Sci., Warsaw(1972).

63. D. VanDulst, Reflexive and superreflexive spaces, Math. Centrum, Tracts 102, Amsterdam, 1978.

64. J. Wells, Differentiable functions on c_0, Bull. Amer. Math. Soc. 75(1969), 117-118.

65. H. Whitney, Analytic extensions of differentiable functions defined on closed sets, Trans. Amer. Math. Soc. 36(1934), 63-80.

66. D. Wulbert, Approximation by C^k functions, Approximation Theory (1973), 217-239.

67. S. Yamamuro, Differential Calculus in topological linear spaces, Springer-Verlag Lecture Notes 373 (1974).

Added in Proof

68. K. John, H. Torunczyk and V. Zizler, Uniformly smooth partitions of unity on superreflexive Banach spaces, Studia Math. LXX, 1981,129-137.

Index

List of Symbols